Springer Theses

Recognizing Outstanding Ph.D. Research

For further volumes:
http://www.springer.com/series/8790

Aims and Scope

The series "Springer Theses" brings together a selection of the very best Ph.D. theses from around the world and across the physical sciences. Nominated and endorsed by two recognized specialists, each published volume has been selected for its scientific excellence and the high impact of its contents for the pertinent field of research. For greater accessibility to non-specialists, the published versions include an extended introduction, as well as a foreword by the student's supervisor explaining the special relevance of the work for the field. As a whole, the series will provide a valuable resource both for newcomers to the research fields described, and for other scientists seeking detailed background information on special questions. Finally, it provides an accredited documentation of the valuable contributions made by today's younger generation of scientists.

Theses are accepted into the series by invited nominated only and must fulfill all of the following criteria

- They must be written in good English.
- The topic of should fall within the confines of Chemistry, Physics and related interdisciplinary fields such as Materials, Nanoscience, Chemical Engineering, Complex Systems and Biophysics.
- The work reported in the thesis must represent a significant scientific advance.
- If the thesis includes previously published material, permission to reproduce this must be gained from the respective copyright holder.
- They must have been examined and passed during the 12 months prior to nomination.
- Each thesis should include a foreword by the supervisor outlining the significance of its content.
- The theses should have a clearly defined structure including and introduction accessible to scientists not expert in that particular field.

Yusuke Ohta

Copper-Catalyzed Multi-Component Reactions

Synthesis of Nitrogen-Containing Polycyclic Compounds

Doctoral Thesis accepted by Kyoto University, Japan

 Springer

Author
Dr. Yusuke Ohta
Graduate School of Pharmaceutical
  Sciences, Kyoto University
Yoshida-shimo-adachi-cho 46-29
Sakyo-ku, Kyoto 606-8501
Japan
e-mail: ssho0108@yahoo.co.jp

Supervisors
Prof. Yoshiji Takemoto and
Prof. Nobutaka Fujii
Graduate School of Pharmaceutical
  Sciences, Kyoto University
Yoshida-shimo-adachi-cho 46-29
Sakyo-ku, Kyoto 606-8501
Japan

ISSN 2190-5053   e-ISSN 2190-5061

ISBN 978-3-642-26701-7   ISBN 978-3-642-15473-7 (eBook)

DOI 10.1007/978-3-642-15473-7

Springer Heidelberg Dordrecht London New York

Cover design: eStudio Clamar, Berlin/Figueres

Printed on acid-free paper

Springer is part of Springer Science+Business Media (www.springer.com)

List of Publications

This study was published in the following papers:

Chapter 1.

1. Direct Synthesis of 2-(Aminomethyl)indoles through Copper(I)-Catalyzed Domino Three-Component Coupling and Cyclization Reactions Hiroaki Ohno, Yusuke Ohta, Shinya Oishi and Nobutaka Fujii *Angew. Chem., Int. Ed.* **2007**, 46, 2295–2298. *Reproduced with permission*

2. Construction of Nitrogen Heterocycles Bearing an Aminomethyl Group by Copper-Catalyzed Domino Three-Component Coupling–Cyclization Yusuke Ohta, Hiroaki Chiba, Shinya Oishi, Nobutaka Fujii and Hiroaki Ohno *J. Org. Chem.* **2009**, *74*, 7052–7058. *Reproduced with permission*

3. Facile Synthesis of 1,2,3,4-Tetrahydro-b-carbolines by One-Pot Domino Three-Component Indole formation and Nucleophilic Cyclization Yusuke Ohta, Shinya Oishi, Nobutaka Fujii and Hiroaki Ohno *Org. Lett.* **2009**, *11*, 1979–1982. *Reproduced with permission*

4. Concise Synthesis of Indole-Fused 1,4-Diazeines through Copper(I)-Catalyzed Domino Three-Component Coupling–Cyclization–*N*-Arylation under Microwave Irradiation Yusuke Ohta, Hiroaki Chiba, Shinya Oishi, Nobutaka Fujii and Hiroaki Ohno *Org. Lett.* **2008**, *10*, 3535–3538. *Reproduced with permission*

Chapter 2.

5. Facile Synthesis of 3-(Aminomethyl)isoquinolines by Copper-Catalyzed Domino Three-Component Coupling and Cyclization Yusuke Ohta, Shinya Oishi, Nobutaka Fujii and Hiroaki Ohno *Chem. Commun.* **2008**, 835–837. *Reproduced with permission*

6. Rapid Access to 3-(Aminomethyl)isoquinoline-Fused Polycyclic Compounds by Copper-Catalyzed Four-Component Coupling, Cascade Cyclization, and Oxidation Yusuke Ohta, Yushi Kubota, Tsuyoshi Watabe, Hiroaki Chiba, Shinya Oishi, Nobutaka Fujii and Hiroaki Ohno *J. Org. Chem.* **2009**, *74*, 6299–6302. *Reproduced with permission*

Supervisor's Foreword

It is a pleasure to introduce Dr. Yusuke Ohta's work for publication in the series *Springer Theses*, as an outstanding original work from one of the world's top universities. Dr. Ohta joined Prof. Fujii's group, Kyoto University, as an undergraduate student from April of 2004. In April 2005, he entered the Graduate School of Pharmaceutical Sciences at Kyoto University, and started his doctoral study with me at the same laboratory.

Multi-component coupling and one-pot reactions have been receiving much attention from many organic chemists because these reactions are useful for green chemistry and atom economy. Dr. Yusuke Ohta developed efficient syntheses of indoles and isoquinolines through multi-component coupling and one-pot reaction catalyzed by copper salt. He reported six outstanding papers in the top journals of Organic Chemistry (*Angewandte Chemie, Organic Letters, the Journal of Organic Chemistry, and Chemical Communications*), some of which were highlightened in Synfact (2009, 7, 726) and Organic Chemistry Portal (2008, September 15).

The thesis results have already inspired further work in progress on efficient synthesis of indoles and isoquinolines, and his findings would contribute to the diversity-oriented synthesis for the drug discovery and facile synthesis of biologically active natural products containing complex structure. I hope his outstanding thesis will contribute to synthetic research of many readers.

Kyoto, April, 2010

Hiroaki Ohno

On behalf of Yoshiji Takemoto
and Nobutaka Fujii

Acknowledgments

The author expresses his sincere and wholehearted appreciation to Professor Nobutaka Fujii (Graduate School of Pharmaceutical Sciences, Kyoto University) for his kind guidance, constructive discussions, and constant encouragement during this study.

The author wishes to express his sincere gratitude to Dr. Hiroaki Ohno (Graduate School of Pharmaceutical Sciences, Kyoto University) and Dr. Shinya Oishi (Graduate School of Pharmaceutical Sciences, Kyoto University) for their thoughtful, tender support, detailed and perceptive comments, and their careful persuing of author's original manuscript.

The author wishes to express his gratitude to Professor Yoshiji Takemoto (Graduate School of Pharmaceutical Sciences, Kyoto University) for his warm encouragement and helpful guidance.

The author also wishes to express his gratitude to Professor Akira Otaka (Institute of Health Biosciendces: IHBS and Graduate School of Pharmaceutical Sciences, the University of Tokushima) for his warm encouragement, constructive discussions, and helpful guidance to the author's first research (Org. Lett. 2006, 8, 467–470).

The author expresses his appreciation to Professor Hirokazu Tamamura (Institute of Biomaterials, and Bioengineering, Tokyo Medical and Dental University), Dr. Hiroyuki Konno (Graduate School of Science and Technology, Yamagata University) for offering helpful comments.

The author is grateful to all the colleagues of Department of Bioorganic Medicinal Chemistry, Graduate School of Pharmaceutical Sciences, Kyoto University, particularly Mr. Hiroaki Chiba and Mr. Yushi Kubota, for their their valuable comments, assistance, and cooperation in various experiments.

The author would like to thank the Japan Society for the Promotion of Science (JSPS) for financial support.

The author is grateful to his parents, Eiji and Reiko Ohta, for their constant source of emotional, moral and financial support throughout his life in Kyoto University. The author is also grateful to his brother, Ryosuke, for the constant encouragement throughout his life in Kyoto University.

Finally, the author thanks his wife, Etsuko, from the bottom of my heart for everything. The author dedicates this work to her.

Contents

Chapter 1
Introduction

One important subject of modern synthetic chemistry is the development of efficient and practical methods for constructing complex heterocyclic structures found in bioactive compounds, natural products, and so on. It is also important to effectively utilize the limited carbon resources minimizing the requisite reagents, solvents, cost, time, separation processes, and wastes [1, 2]. The multi-component reaction (MCR) [3–6], represented by Ugi's four-component coupling (Scheme 1) [3, 7], is well recognized as a powerful approach toward these ends. MCR is a convergent reaction in which one product is yielded from three or more materials, and can produce a variety of compounds if only each material is changed. MCRs provide easy access to combinatorial chemistry, diversity-oriented synthesis, and high throughput screening saving carbon resources. A catalytic domino reaction [2, 8–10] including MCR would be more attractive to achieve these goals since it can make it possible to form multiple bonds.

Since the indole nucleus is a prominent structural motif found in numerous natural products and synthetic compounds with vital biological activities, considerable attention has been directed toward general, flexible, and selective synthetic methods for highly functionalized indole derivatives [11, 12]. Among the functionalized indoles, 2-(aminomethyl)indole motif represents the key structures that exist in several biologically active compounds [13–25] including calindol (Fig. 1) [26–28]. Most of the synthetic routes to 2-(aminomethyl)indoles rely upon the functionalized indoles such as indole-2-carboxylic acid or its derivatives as the starting materials [26–30], which limit the structure of the target molecules that can be readily synthesized.

The isoquinoline scaffold can be found in a wide variety of biologically active natural and synthetic compounds [31–38]. Particularly, isoquinolines having an additional nitrogen atom tethered by one carbon at the 3-position, including such isoquinoline alkaloids as quinocarcine [39–42], ecteinascidins 597 and 583 [43, 44], and 3-(2-pyridinyl)isoquinolines [45–47] constitute an important class of compounds with important biological activities (Fig. 2).

Y. Ohta, *Copper-Catalyzed Multi-Component Reactions*, Springer Theses, DOI: 10.1007/978-3-642-15473-7_1, © Springer-Verlag Berlin Heidelberg 2011

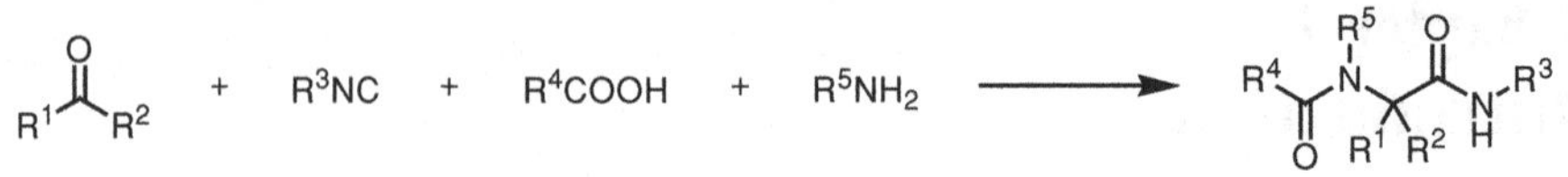

Scheme 1 Ugi's four-component coupling reaction

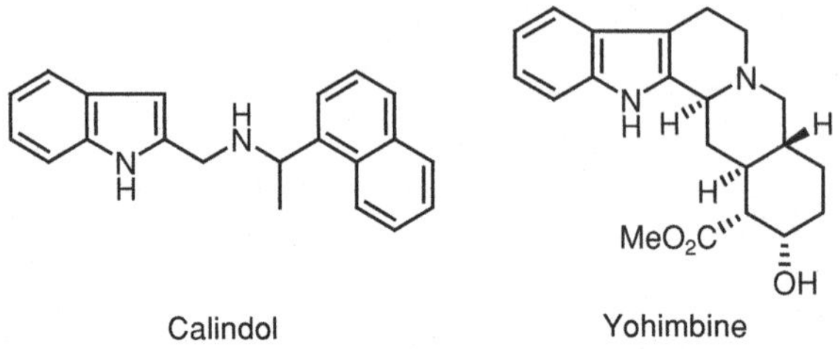

Fig. 1 Compounds containing a 2-(Aminomethyl)indole Motif

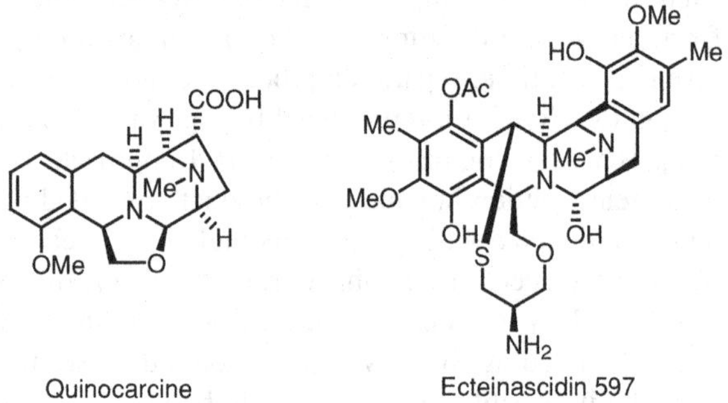

Calindol

Yohimbine

Fig. 2 Natural products containing 3-(Aminomethyl)isoquinoline Motif

Quinocarcine

Ecteinascidin 597

Synthesis of indole derivatives by a catalytic domino three-component reaction including Sonogashira-type cross-coupling of dihalobenzenes [48, 49] or haloanilines [50–52] has been recently accomplished [53, 54]. Ackermann reported synthesis of indoles through Sonogashira coupling of 2-chloro-1-iodobenzene and a terminal alkyne followed by N-arylation and intramolecular hydroamination (Eq. 1) [48, 49]. Alami synthesized 2-(aminomethyl)indoles by S_N2 reaction of a secondary amine with propargylic bromide, Sonogashira coupling with 2-iodoaniline, and hydroamination (Eq. 2) [50]. Senanayake succeeded in construction of 2,3-disubstituted indoles through Sonogashira coupling, insertion of aryl palladium halide to alkyne moiety, and C–N bond formation (Eq. 3) [51].

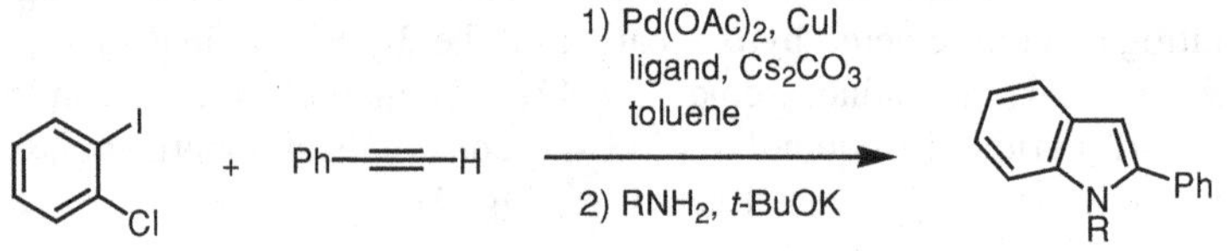

$$\text{1) } Pd(OAc)_2, CuI$$
$$\text{ligand, } Cs_2CO_3$$
$$\text{toluene}$$
$$\text{2) } RNH_2, t\text{-BuOK}$$

(Eq.1)

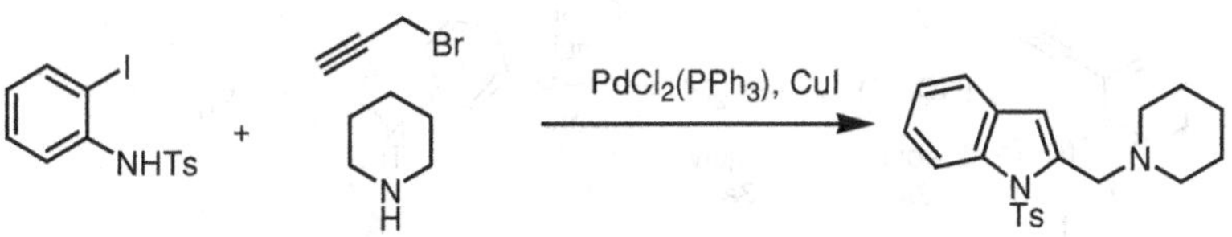

(Eq.2)

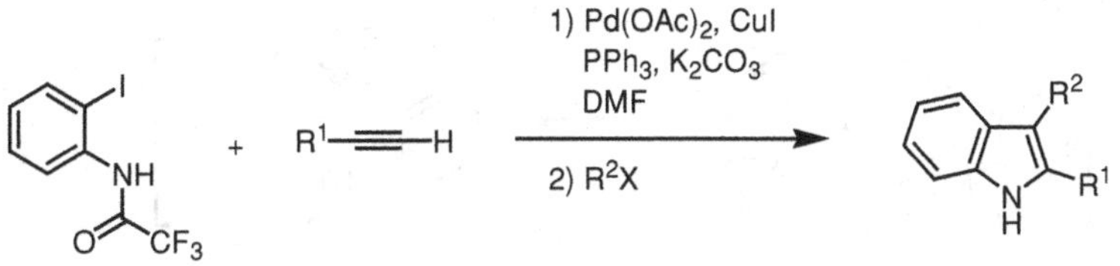

(Eq.3)

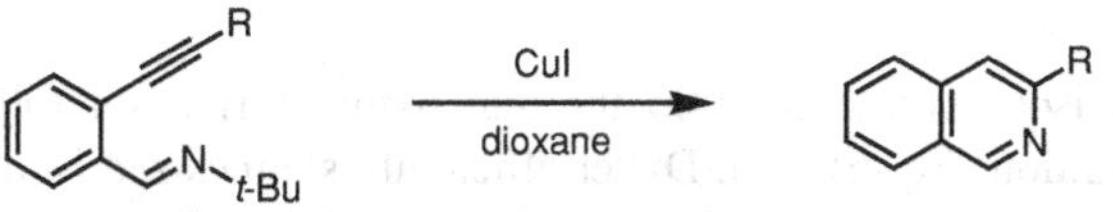

(Eq.4)

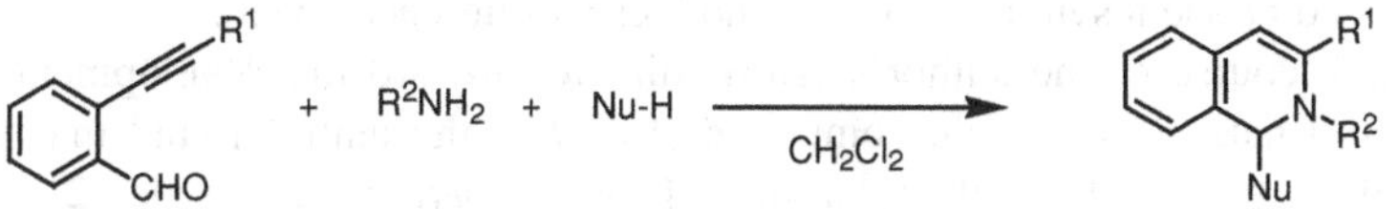

(Eq.5)

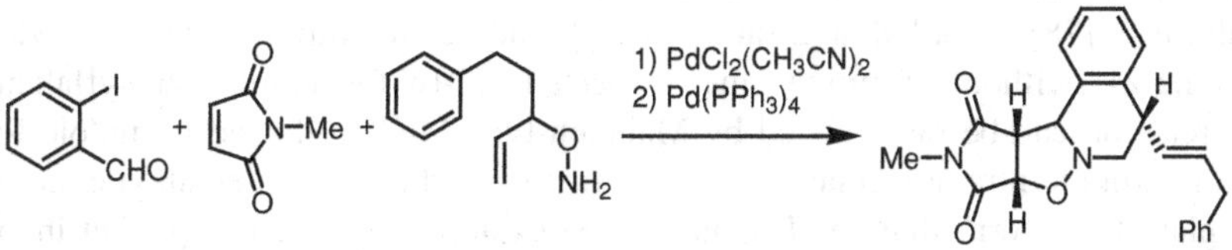

(Eq.6)

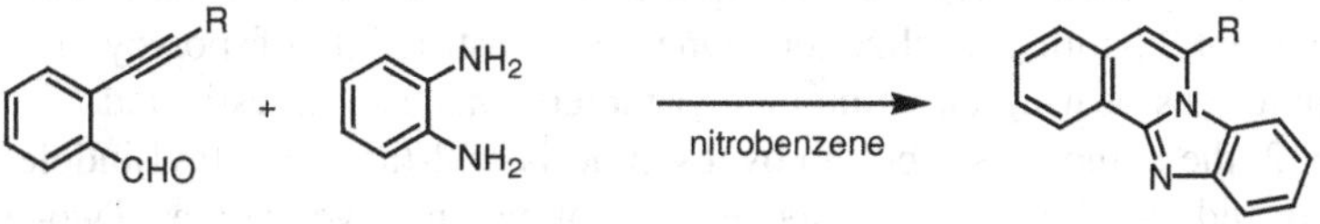

(Eq.7)

Larock developed a powerful approach to isoquinolines which involves copper-catalyzed hydroamination of N-tert-butyl-2-(1-alkynyl)benzaldimine accompanied by elimination of tert-butyl group (Eq. 4) [55–59]. Asao and Yamamoto reported a novel synthesis of 1,2,3-trisubstituted isoquinolines through attack of a carbon nucleophile to the carbon–nitrogen double bond of N-alkyl-2-(1-alkynyl)benzaldimine and simultaneous hydroamination catalyzed by transition metal [60]. They also achieved isoquinoline synthesis by transition metal-free three-component coupling (Eq. 5) [61]. Takemoto and Yanada reported a related isoquinoline formation by a catalysis of carbophilic Lewis acids such as indium(III), Ni(II), or Au(I)/Ag(I) [62, 63]. Oikawa succeeded in palladium-catalyzed three-component

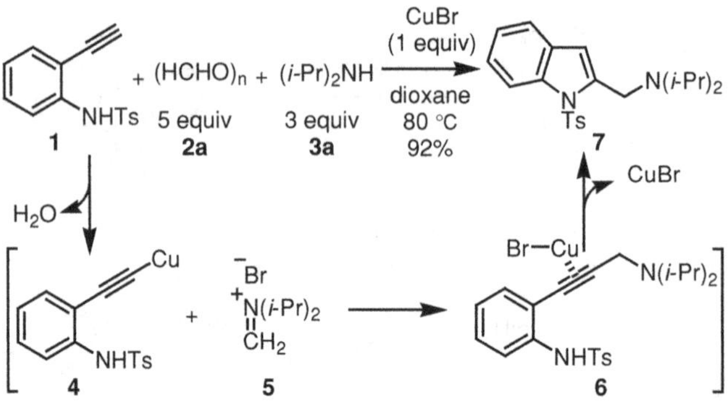

Scheme 2 Domino three-component coupling–cyclization

construction of isoquinoline scaffold through oxime formation followed by 1,3-dipolar cycloaddition (Eq. 6) [64]. Dyker efficiently synthesized isoquinoline-fused polycyclic compounds using phenylenediamine (Eq. 7) [65]. Despite these successful studies, four-component synthesis of isoquinolines was unprecedented.

During the course of the author's efforts directed toward the development of useful transformations of allenic compounds [66–77], the author found that the reaction of *N*-tosylated 2-ethynylaniline **1** with paraformaldehyde **2** and diisopropylamine **3** in dioxane in the presence of copper(I) bromide (Crabbé conditions) [78] afforded a 2-(aminomethyl)indole derivative **7** in 92% yield (Scheme 2) without forming the expected [2-(*N*-tosylamino)phenyl]allene. This reaction can be rationalized by Mannich-type MCR followed by indole formation through intramolecular hydroamination toward the activated alkyne moiety of a plausible intermediate **6**. This is the first example of three-component indole formation without producing stoichiometric amount of salts as byproducts.

In this study, the author examined an atom-economical and diversity-oriented synthesis of 2-(aminomethyl)indoles/isoquinolines by copper-catalyzed domino multi-component coupling–cyclization. One-pot construction of polycyclic indoles/isoquinolines bearing an aminomethyl moiety was also investigated.

In Chap. 2, the author describes a novel synthesis of 2-(aminomethyl)indole by copper-catalyzed domino three-component coupling and cyclization. Two-step construction of polycyclic indoles by combination with palladium-catalyzed C–H functionalization at the indole C-3 position, scope and limitation of the asymmetric three-component indole formation, and synthesis of benzo[*e*][1, 2]thiazine derivatives and indene-1,1-dicarboxylate, are also presented in this section.

In Chap. 3, the author describes two direct routes to 1,2,3,4-tetrahydro-*β*-carboline derivatives by a copper-catalyzed one-pot three-component coupling–indole formation–nucleophilic cyclization at the 3-position of indole.

In Chap. 4, the author describes a direct access to indole-fused tetracyclic compounds containing a 1,4-diazepine framework by copper-catalyzed domino three-component coupling, cyclization, and *N*-arylation, which involve the formation of one carbon–carbon bond and three carbon–nitrogen bonds.

In Chap. 5, the author describes copper-catalyzed domino four-component coupling–cyclization reaction for diversity-oriented synthesis of 3-(aminomethyl)-isoquinolines.

In Chap. 6, the author describes a novel approach to 3-(aminomethyl)isoquinolinefused polycyclic compounds utilizing four-component coupling and cascade cyclization in the presence of a copper catalyst.

References

1. Trost BM (2002) Acc Chem Res 35:696
2. Nicolaou KC, Montagnon T, Snyder SA (2003) Chem Commun 551
3. Dömling A, Ugi I (2000) Angew Chem Int Ed 39:3168
4. Dömling A (2006) Chem Rev 106:17
5. Tejedor D, García-Tellado F (2007) Chem Soc Rev 36:484
6. D'Souza DM, Müller TJJ (2007) Chem Soc Rev 36:1095
7. Ugi I (1962) Angew Chem Int Ed 1:8
8. Malacria M (1996) Chem Rev 96:289
9. Nicolaou KC, Edmonds DJ, Bulger PG (2006) Angew Chem Int Ed 45:7134
10. Enders D, Grondal C, Hüttl MRM (2007) Angew Chem Int Ed 46:1570
11. Humphrey GR, Kuethe JT (2006) Chem Rev 106:2875
12. Cacchi S, Fabrizi G (2005) Chem Rev 105:2873
13. Bosch J, Bennasar M-L (1995) Synlett 587
14. Saxton JE (1997) Nat Prod Rep 14:559
15. Leonard J (1999) Nat Prod Rep 16:319
16. Lobo AM, Prabhakar SJ (2002) Heterocycl Chem 39:429
17. Takayama H (2005) Chem Pharm Bull 52:916
18. Takayama H, Kitajima M, Kogure N (2005) Curr Org Chem 9:1445
19. Lewis SE (2006) Tetrahedron 62:8655
20. Morón JA, Campillo M, Perez V, Unzeta M, Pardo L (2000) J Med Chem 43:1684
21. Spadoni G, Balsamini C, Diamantini G, Tontini A, Tarzia G (2001) J Med Chem 44:2900
22. Rivara S, Mor M, Silva C, Zuliani V, Vacondio F, Spadoni G, Bedini A, Tarzia G, Lucini V, Pannacci M, Fraschini F, Plazzi PV (2003) J Med Chem 46:1429
23. Stewart AO, Cowart MD, Moreland RB, Latshaw SP, Matulenko MA, Bhatia PA, Wang X, Daanen JF, Nelson SL, Terranova MA, Namovic MT, Donnelly-Roberts DL, Miller LN, Nakane M, Sullivan JP, Brioni JD (2004) J Med Chem 47:2348
24. Rivara S, Lorenzi S, Mor M, Plazzi PV, Spadoni G, Bedini A, Tarzia G (2005) J Med Chem 48:4049
25. Brands M, Ergüden J-K, Hashimoto K, Heimbach D, Schröder C, Siegel S, Stasch J-P, Weigand S (2005) Bioorg Med Chem Lett 15:4201
26. Kessler A, Faure H, Petrel C, Ruat M, Dauban P, Dodd RH (2004) Bioorg Med Chem Lett 14:3345
27. Petrel C, Kessler A, Dauban P, Dodd RH, Rognan D, Ruat M (2004) J Biol Chem 279:18990
28. Ray K, Tisdale J, Dodd RH, Dauban P, Ruat M, Northup JK (2005) J Biol Chem 280:37013
29. Ambrogio I, Cacchi S, Fabrizi G (2006) Org Lett 8:2083
30. Pedras MSC, Suchy M, Ahiahonu PWK (2006) Org Biomol Chem 4:691
31. Scott JD, Williams RM (2002) Chem Rev 102:1669
32. Chrzanowska M, Rozwadowska MD (2004) Chem Rev 104:3341
33. Bermejo A, Andreu I, Suvire F, Leonce S, Caignard DH, Renard P, Pierré A, Enriz RD, Cortes E, Cabedo N (2002) J Med Chem 45:5058
34. Morrel A, Antony S, Kohlhagen G, Pommier Y, Cushman MJ (2006) J Med Chem 49:7740

35. Bringmann G, Dreyer M, Faber JH, Dalsgaard PW, Stærk D, Jaroszewski JW, Ndangalasi H, Mbago F, Brun R, Christensen SB (2004) J Nat Prod 67:743
36. Graulich A, Mercier F, Scuvée-Moreau J, Seutin V, Liégeois JF (2005) Bioorg Med Chem 13:1201
37. Chen YH, Zhang YH, Zhang HJ, Liu DZ, Gu M, Li JY, Wu F, Zhu XZ, Li J, Nan FJ (2006) J Med Chem 49:1613
38. Bringmann G, Mutanyatta-Comar J, Greb M, Rüdenauer S, Noll TF, Irmer A (2007) Tetrahedron 63:1755
39. Tomita F, Takahashi K, Shimizu K (1983) J Antibiot 36:463
40. Takahashi K, Tomita F (1983) J Antibiot 36:468
41. Fukuyama T, Nunes JJ(1988) J Am Chem Soc 110:5196
42. Kwon S, Myers AG (2005) J Am Chem Soc 127:16796
43. Sakai R, Jares-Erijman EA, Manzanares I, Elipe MVS, Rinehart KL (1996) J Am Chem Soc 118:9017
44. Chen J, Chen X, Willot M, Zhu J (2006) Angew Chem Int Ed 45:8028
45. de Zwart MAH, van der Goot H, Timmerman H (1989) J Med Chem 32:487
46. van Muijlwijk-Koezen JE, Timmerman H, Link R, van der Goot H, IJzerman AP (1998) J Med Chem 41:3987
47. van Muijlwijk-Koezen JE, Timmerman H, Link R, van der Goot H, IJzerman AP (1998) J Med Chem 41:3994
48. Ackermann L (2005) Org Lett 7:439
49. Kaspar LT, Ackermann L (2005) Tetrahedron 61:11311
50. Olivi N, Spruyt P, Peyrat J-F, Alami M, Brion J-D (2004) Tetrahedron Lett 45:2607
51. Lu BZ, Zhao W, Wei H-X, Dufour M, Farina V, Senanayake CH (2006) Org Lett 8:3271
52. Sanz R, Guilarte V, Pérez A (2009) Tetrahedron Lett 50:4423
53. Cacchi S, Fabrizi G, Parisi LM (2003) Org Lett 5:3843
54. McLaughlin M, Palucki M, Davies IW (2006) Org Lett 8:3307
55. Roesh KR, Larock RC (2002) J Org Chem 67:86
56. Roesh KR, Larock RC (1998) J Org Chem 63:5306
57. Huang Q, Hunter JA, Larock RC (2001) Org Lett 3:2973
58. Huang Q, Hunter JA, Larock RC (2002) J Org Chem 67:3437
59. Zhang H, Larock RC (2002) Tetrahedron Lett 43:1359
60. Asao N, Yudha SS, Nogami T, Yamamoto Y (2005) Angew Chem Int Ed 44:5526
61. Asao N, Iso K, Yudha SS (2006) Org Lett 8:4149
62. Yanada R, Obika S, Kono H, Takemoto Y (2006) Angew Chem Int Ed 45:3822
63. Obika S, Kono H, Yasui Y, Yanada R, Takemoto Y (2007) J Org Chem 72:4462
64. Oikawa M, Takeda Y, Naito S, Hashizume D, Koshino H, Sasaki M (2007) Tetrahedron Lett 48:4255
65. Dyker G, Stirner W, Henkel G (2000) Eur J Org Chem 1433
66. Ohno H, Hamaguchi H, Ohata M, Tanaka T (2003) Angew Chem Int Ed 42:1749
67. Ohno H, Miyamura K, Takeoka Y, Tanaka T (2003) Angew Chem Int Ed 42:2647
68. Ohno H, Hamaguchi H, Ohata M, Kosaka S, Tanaka T (2004) J Am Chem Soc 126:8744
69. Hamaguchi H, Kosaka S, Ohno H, Tanaka T (2005) Angew Chem Int Ed 44:1513
70. Ohno H, Mizutani T, Kadoh Y, Miyamura K, Tanaka T (2005) Angew Chem Int Ed 44:5113
71. Ohno H, Kadoh Y, Fujii N, Tanaka T (2006) Org Lett 8:947
72. Ohno H, Aso A, Kadoh Y, Fujii N, Tanaka T (2007) Angew Chem Int Ed 46:6325
73. Watanabe T, Oishi S, Fujii N, Ohno H (2007) Org Lett 9:4821
74. Okano A, Mizutani T, Oishi S, Tanaka T, Ohno H, Fujii N (2008) Chem Commun 3534
75. Inuki S, Oishi S, Fujii N, Ohno H (2008) Org Lett 10:5239
76. Ohno H (2005) Chem Pharm Bull 53:1211
77. Ohno H (2005) Yakugaku Zasshi 125:899
78. Searles S, Nassim Y, Li B, Lopes M-TR, Tran PT, Crabbé P (1984) J Chem Soc, Perkin Trans 1:747

Part I
Synthesis of Indole Derivatives

Chapter 2
Construction of 2-(Aminomethyl)indoles Through Copper-Catalyzed Domino Three-Component Coupling and Cyclization

2.1 Introduction

As described in preface, the author found that the reaction of N-tosylated 2-ethynylaniline **1a** with paraformaldehyde **2a** and diisopropylamine **3a** in dioxane in the presence of copper(I) bromide afforded a 2-(aminomethyl)indole derivative **7a** in 92% yield (Scheme 1). This reaction can be rationalized by Mannich-type MCR followed by indole formation through intramolecular hydroamination toward the activated alkyne moiety of a plausible intermediate **6**. Actually, the reaction of the identically prepared propargyl amine **8** with CuBr (5 mol.%) gave the expected indole **7b** in quantitative yield (Scheme 2).

To improve the original reaction conditions using a stoichiometric amount of CuBr and 3 equiv of $(i\text{-Pr})_2$NH (Scheme 1), the initial attempt was made by reacting with N-tosyl-2-ethynylaniline **1a**, paraformaldehyde **2a** (2 equiv), piperidine **3b** (1.1 equiv), and CuBr (100 mol.%) in the presence of Et$_3$N (2 equiv) which would decrease the loading of piperidine (Table 1, entry 1).[1] The reaction proceeded rapidly to give the desired 2-(aminomethyl)indole **7b** in 71% yield. While use of a catalytic amount of CuBr (10 or 1 mol.%) with respect to **1a** increased the yield of **7b** (entries 2 and 3), the reaction without CuBr led to the recovery of **1a**. The reaction in the absence of Et$_3$N also showed efficient conversion into **7b** (entry 4). This result can be explained by the plausible reaction mechanism depicted in Scheme 1, in which the sulfonamide proton is presumably transferred to the 3-position of indole. This step could be mediated by piperidine or the basic substituent in the product and/or intermediate. The decreased use of **2a** also produced the desired indole **7b**, although a prolonged reaction time (1–12 h) was necessary (entries 5 and 6). Use of CuBr$_2$, CuCl, or CuI as the catalyst was also tolerated in this three-component indole formation (entries 7–9).

---

[1] The author considered decreasing of the amount of amine component is important and economical especially when using more valuable amines such as **11** (Scheme 4).

Y. Ohta, *Copper-Catalyzed Multi-Component Reactions*, Springer Theses, DOI: 10.1007/978-3-642-15473-7_2, © Springer-Verlag Berlin Heidelberg 2011

2.1.1 Synthesis of 2-(Aminomethyl)indoles Using Several Amines and Aldehydes

Next, the author examined the scope of the 2-(aminomethyl)indole formation with various symmetrical secondary amines (Table 2) under the optimized conditions (Table 1, entry 4). The reaction of 2-ethynylaniline **1a** with bulky diisopropyl-amine **3a** (1.1 equiv) and paraformaldehyde **2a** (2 equiv) in the presence of CuBr (1 mol.%) gave the expected indole derivatives **7a** in 81% yield (entry 1). Pyr-rolidine **3c** also showed efficient conversion into the corresponding indoles **7c** (entry 3). The use of volatile diethylamine **3d** successfully afforded **7d**, although 2 equiv of Et_2NH were needed (entry 4). Secondary amines containing removable allyl and benzyl groups **3e** and **3f**, respectively, were also acceptable as amine components when the reactions were conducted with a prolonged reaction time (entries 5 and 6).[2]

The author also investigated the three-component synthesis of 2-(amino-methyl)indoles using various aldehyde components (Table 3). The reaction of 2-ethynylaniline **1a** with butanal **2b** and piperidine **3b** in the presence of CuBr efficiently gave the indole **7g** bearing a branched substituent in an excellent yield (quant., entry 1). The bulky *i*-butyraldehyde **2c** required an elevated reaction temperature and prolonged reaction time leading to a slightly decreased yield of **7h** (77%, entry 2). Benzaldehyde **2d** was tolerated for this indole formation (entry 3). Similarly, use of a variety of substituted aryl aldehydes afforded the desired indoles **7j–7l** in good yields (entries 4–6).[3]

The author expected that a reaction with a chiral ligand which coordinates to a copper atom could produce optically active 2-(aminomethyl)indoles. Knochel recently developed a novel asymmetric synthesis of chiral propargylamines with excellent ee values through a copper-catalyzed asymmetric Mannich-type reaction of alkynes with an aldehyde and a secondary amine using QUINAP as a chiral ligand (up to 98% ee) [1–3]. Carreira reported the similar synthesis of propargylic amine in up to 99% ee with PINAP [4, 5]. The author initially examined the

---

[2] When benzylamine was used instead of a secondary amine, dimeric compound **18** was produced in 82% yield (100 °C, 3 h, then reflux, 1 h).

18

[3] When acetone was used instead of an aldehyde, Mannich-type reaction did not proceed and compound **19** was produced.

19

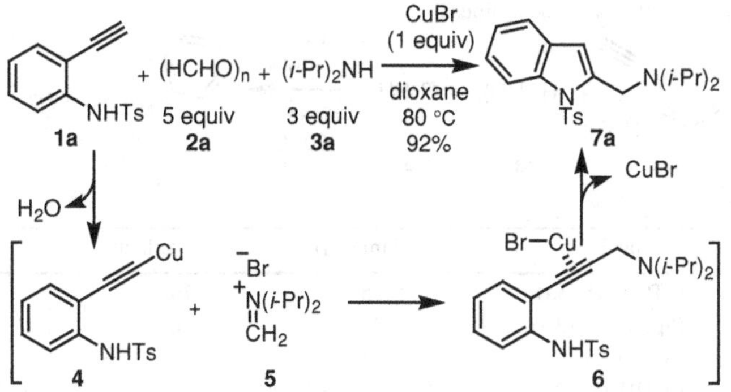

Scheme 1 Domino three-component coupling–cyclization

Scheme 2 Indole formation from the proposed intermediate **8**

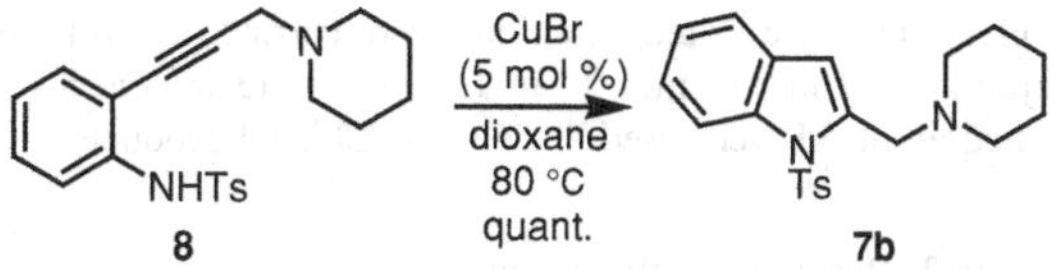

Table 1 Optimization of reaction conditions using ethynylaniline **1a** and piperidine **3b**

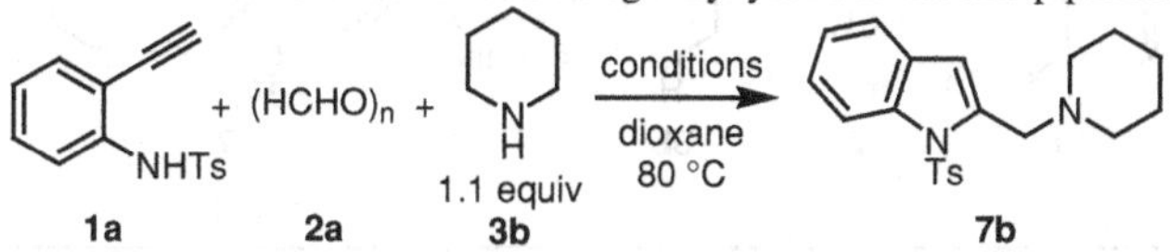

Entry	CuX (mol.%)	(HCHO)$_n$ (equiv)	Additive (equiv)	Time (h)	Yield[a] (%)
1	CuBr (100)	2.0	Et$_3$N (2)	0.25	71
2	CuBr (10)	2.0	Et$_3$N (2)	0.25	84
3[b]	CuBr (1)	2.0	Et$_3$N (2)	0.25	92
4	CuBr (1)	2.0	–	0.25	87
5	CuBr (1)	1.5	–	1	75
6	CuBr (1)	1.1	–	12	70
7	CuBr$_2$ (1)	2.0	–	0.25	79
8	CuCl (1)	2.0	–	0.25	87
9	CuI (1)	2.0	–	0.25	83

Unless otherwise stated, reaction was carried out with **1a** (0.18 mmol, 1 equiv), **2a** (equiv shown), **3b** (1.1 equiv), and a copper salt (catalyst amount shown) in 1,4-dioxane (3 mL) at 80 °C
[a] Yields of isolated products. [b] The reaction was conducted on 1.25 mmol scale

asymmetric three-component construction of the 2-(aminomethyl)indole motif with *n*-butyraldehyde **2b** in dioxane in the presence of CuBr (5 mol.%) and QUINAP (5.5 mol.%) (Table 4). The reaction proceeded smoothly even at rt to give the desired **7g** in a quantitative yield but with only 47% ee (entry 1). It was reported that the copper-catalyzed Mannich reaction of alkynes in the presence of (*R*)-QUINAP gave (*S*)-propargylamines, while the reaction with (*S*)-PINAP gave

Table 2 Reactions with various amines

Entry	Amine **3**	Time (h)	Product	Yield (%)[b]
1	(*i*-Pr)$_2$NH (**3a**)	0.25	**7a**	81
2	Piperidine (**3b**)	0.25	**7b**	87
3	Pyrrolidine (**3c**)	0.25	**7c**	89
4	Et$_2$NH (**3d**)[a]	0.25	**7d**	89
5	(allyl)$_2$NH (**3e**)	0.5	**7e**	78
6	Bn$_2$NH (**3f**)	2	**7f**	78

Unless otherwise stated, reactions were carried out with **1a** (0.18 mmol), **2a** (2.0 equiv), **3** (1.1 equiv), and CuBr (1 mol.%) in 1,4-dioxane (3 mL) at 80 °C
[a] 2 equiv of **3d** were used, [b] yields of isolated products

Table 3 Reactions with various aldehydes

Entry	Aldehyde **2**	Conditions	Product yield (%)[a]
1	*n*-PrCHO (**2b**)	80 °C 0.25 h	**7g** (R = *n*-Pr) quant.
2	*i*-PrCHO (**2c**)	Reflux 3 h	**7h** (R = *i*-Pr) 77
3	PhCHO (**2d**)	Reflux 10 h	**7i** (R = Ph) 70
4	(4-CO$_2$Me)C$_6$H$_4$CHO (**2e**)	Reflux 3 h	**7j** [R = (4-CO$_2$Me)C$_6$H$_4$] 76
5	(4-Me)C$_6$H$_4$CHO (**2f**)	Reflux 3 h	**7k** [R = (4-Me)C$_6$H$_4$] 85
6	(2-Br)C$_6$H$_4$CHO (**2g**)	Reflux 4 h	**7l** [R = (2-Br)C$_6$H$_4$] 65

Reactions were carried out with **1a** (0.18 mmol), **2** (2.0 equiv), **3b** (1.1 equiv), and CuBr (1 mol.%) in 1,4-dioxane (1.5 mL) at 80 °C
[a] Yields of isolated products

the corresponding (*R*)-isomers, see Refs. 1–5. Screening of the reaction solvent did not improve the asymmetric induction (entries 2–4). When the reaction was carried out with PINAP in dioxane, **7g** was obtained with a slightly higher ee (59%, ee, entry 5). Use of PINAP in benzene gave the most promising result (63% ee), although a prolonged reaction time was necessary (entry 7). These results suggest that 2-ethynylaniline **1a** is a less effective alkyne component for an asymmetric Mannich reaction. Knochel and Carreira reported that phenylacetylene is a good component for enantioselective synthesis of propargylic amine using QUINAP or PINAP, see Refs. 1–5.

Table 4 Asymmetric synthesis of 2-(aminomethyl)indoles

Entry	Ligand	Solvent	Conditions	Yield[a] (%)	Product (% ee)[b]
1	(R)-QUINAP	Dioxane	rt, 24 h	quant.	(+)-**7g** (47)
2	(R)-QUINAP	THF	rt, 72 h	86	(+)-**7g** (30)
3	(R)-QUINAP	benzene	rt, 72 h	86	(+)-**7g** (43)
4	(R)-QUINAP	Toluene	rt, 72 h	94	(+)-**7g** (22)
5	(S)-PINAP	Dioxane	rt, 10 h	quant.	(−)-**7g** (59)
6	(S)-PINAP	Toluene	rt, 120 h	93	(−)-**7g** (56)
7	(S)-PINAP	Benzene	rt, 120 h	quant.	(−)-**7g** (63)

(R)-QUINAP (S)-PINAP

Reactions were carried out with **1a**, CuBr (5 mol.%), ligand (5.5 mol.%) in solvent (2 mL)
[a] Yields of isolated products, [b] determined by chiral HPLC (CHIRALCEL OD-H)

2.1.2 Synthesis of Substituted 2-(Aminomethyl)indoles Using Various Ethynylanilines and Secondary Amines

Various substituted 2-ethynylanilines and asymmetrical secondary amines were then applied to the domino three-component coupling–cyclization (Table 5). 2-Ethynylanilines **1b** and **1c** substituted by electron-withdrawing trifluoromethyl or methoxycarbonyl group at the para position to the amino group were reacted with paraformaldehyde **2a** and dibenzylamine **3f** in the presence of CuBr (1 mol.%) to yield indoles **7m** (90% yield) and **7n** (91% yield), respectively (entries 1 and 2). Ethynylaniline **1d** bearing an electron-donating methyl group at the para position to the amino group also showed efficient compatibility leading to the corresponding indole **7o**. The reaction using 2-ethynylanilines **1e** and **1f** containing an electron-withdrawing group such as a trifluoromethyl or methoxycarbonyl group at the meta position were similarly converted into the corresponding indoles **7p** (61% yield) and **7q** (79% yield), respectively (entries 4, 5). The asymmetrical 2-bromoallylamine **3g** and 2-bromobenzylamine **3h** were also applicable to this indole formation using various 2-ethynylanilines (entries 6–11), although Et$_3$N was necessary for the cyclization step when using 2-ethynylanilines **1a** and **1d**.

2.1.3 Construction of Polycyclic Indoles by Palladium-Catalyzed C–H Functionalization

A polycyclic indole motif is an important core framework which is widely found in biologically active compounds. For biologically active polycyclic indoles having a 2-(aminomethyl) moiety, see [6–10]. Therefore, development of a convenient and reliable method for the construction of these frameworks is strongly required. For recent synthesis of polycyclic indoles, see [11–13]. The author expected that the present synthesis of 2-(aminomethyl)indoles via domino three-component coupling–cyclization would bring about an extremely useful synthetic route to this class of compounds. Thus, the author surveyed the construction of polycyclic indole skeletons by three-component indole formation followed by palladium-catalyzed C–H functionalization at the C-3 position of indoles. First, 2-(amino-methyl)indole **7r** synthesized by the three-component indole formation (Table 5, entry 6) was subjected to Pd(OAc)$_2$ (10 mol.%), PPh$_3$ (20 mol.%), and CsOAc (2 equiv) in DMF (Table 6, entry 1). The reaction proceeded cleanly to afford tetrahydropyridine-fused indole **9a** in 47% yield. When DMA was used as the reaction solvent, a higher yield of **9a** was observed (65%, entry 2). Further investigation of the palladium catalyst, ligand, and base (entries 3–5) revealed that the conditions shown in entry 2 were most effective.

Encouraged by this result, the author investigated the reaction with several 2-(aminomethyl)indoles containing an electron-withdrawing and -donating group to obtain variously substituted tetrahydropyridine-fused indoles **9b–f** in moderate to good yields (Table 7).

The author next examined construction of polycyclic indoles by palladium-catalyzed C–H arylation using 2-(aminomethyl)indole **7x**, which was prepared from ethynylaniline **1a** and amine **3h** (Table 5, entry 11). By treatment with 20 mol.% of Pd(OAc)$_2$ and 40 mol.% of PPh$_3$, dihydrobenzazepine-fused indole **10** was efficiently obtained in 80% yield over 2 steps (Scheme 3). One-pot three-component indole formation/Pd-catalyzed C–H arylation also provided polycyclic indole **10** in 84% yield from **1a**.

2.1.4 Synthetic Application to Calindol, Benzo[e][1,2]thiazines, and Indene

Calindol (**13**), which contains a 2-(aminomethyl)indole motif, is a positive modulator of the human Ca^{2+} receptor showing a calcimimetic activity [1–3]. This compound could be easily synthesized using this domino three-component indole formation (Scheme 4). As the author expected, the reaction of 2-ethynylaniline **1a** with paraformaldehyde **2a** and 1-(1-naphthyl)ethylamine **11** in presence of CuBr directly produced a protected calindol **12**. The allyl and tosyl groups on the nitrogen atoms of **12** were easily removed by successive treatment with Pd(PPh$_3$)$_4$ (2 mol.%)/NDMBA and TBAF [14] to give calindol **13** in 90% yield over 2 steps.

Table 5 Synthesis of variously substituted 2-(aminomethyl)indoles

Entry	2-ethynylaniline	Amine	Conditions	Product (yield[c])
		Bn₂NH		
1	**1b** (R = CF₃)	**3f**	80 °C, 3 h	**7m** (R = CF₃, 90%)
2	**1c** (R = CO₂Me)		80 °C, 5 h	**7n** (R = CO₂Me, 91%)
3	**1d** (R = Me)		80 °C, 5 h, then reflux, 1 h	**7o** (R = Me, 78 %)
4	**1e** (R = CF₃)		80 °C, 3 h	**7p** (R = CF₃, 61%)
5	**1f** (R = CO₂Me)		80 °C, 5 h	**7q** (R = CO₂Me, 79%)
6[a]	**1a** (R = H)	**3g**	80 °C, 3 h, then reflux,[b] 1 h	**7r** (R = H, 98%)
7[a]	**1b** (R = CF₃)		80 °C, 3 h	**7s** (R = CF₃, 91%)
8[a]	**1c** (R = CO₂Me)		80 °C, 3 h	**7t** (R = CO₂Me, 98%)
9[a]	**1d** (R = Me)		80 °C, 3 h, then reflux,[b] 3 h	**7u** (R = Me, 98%)
10[a]	**1e** (R = CF₃)	**3g**	80 °C, 3 h, then reflux, 1 h	**7v** (R = CF₃, 94%)
11[a]	**1f** (R = CO₂Me)		80 °C, 3 h, then reflux, 1.5 h	**7w** (R = CO₂Me, 99%)
12	**1a**	**3h**	80 °C, 3 h, then reflux, 1 h	**7x** (80%)

Unless otherwise stated, reactions were carried out with **1** (0.18 mmol), **2a** (2.0 equiv), **3** (1.1 equiv), and CuBr (1 mol.%) in 1,4-dioxane (3 mL) [a] 0.37 mmol scale, [b] 4 equiv of Et₃N were added before reflux, [c] yields of isolated products

The author next envisioned the preparation of benzothiazine-1,1-dioxide derivatives **15** through domino MCR and cyclization. Since benzo[*e*][1,2]thiazine-1,1-dioxides are widely found in biologically active compounds including nonsteroidal anti-inflammatory drugs (NSAIDs) [15–22], various approaches to construct this structure have been reported [23–33]. The author expected that the use of such a sulfonamide as **14**, an aldehyde, and a secondary amine in the presence of a copper catalyst would bring about a Mannich-type reaction followed by 6-*endo-dig* cyclization (related synthesis of thiazines has been already reported, see [34, 35]) to afford a benzo[*e*][1,2]thiazine **15**. The reaction of *N*-methyl and *N*-ethylsulfonamides **14a** and **14b** under standard conditions gave the desired benzothiazines **15a** and **15b**, respectively, but in low yields (34 and 37%, respectively, entries 1 and 2, Table 8). Considering that acidity of the amide proton in **14a** and

Table 6 Palladium-catalyzed C–H olefination

Entry	Catalyst	Ligand	Base	Solvent	Yield (%)[a]
1	Pd(OAc)$_2$	PPh$_3$	CsOAc	DMF	47
2	Pd(OAc)$_2$	PPh$_3$	CsOAc	DMA	65
3	Pd(PPh$_3$)$_4$	–	CsOAc	DMA	7
4	Pd(OAc)$_2$	PPh$_3$	KOAc	DMA	35
5	Pd(OAc)$_2$	dppm	CsOAc	DMA	32

Reactions were carried out with 2-(aminomethyl)indole **7r**, palladium catalyst (10 mol.%), ligand (20 mol.%), and base (2 equiv) in solvent (2 mL) at 100 °C for 0.5 h
[a] Yields of isolated products

14b would be insufficient for the cyclization step, the author next examined the reaction of sulfonanilide derivatives bearing a related structure to 2-ethynylanilines **1**. As the author expected, the reaction of sulfonanilide **14c** gave the benzothiazine **15c** in high yield (90%, entry 3). Other sufonanilides **14d–14f** were also good reactants in this three-component thiazine synthesis (entries 4–6).

Finally, the author investigated the synthesis of 2-(aminomethyl)indene-1,1-dicarboxylate **17** using this domino Mannich-type reaction/cyclization strategy (Table 9). Disappointingly, the reaction of malonate derivative **16** with (HCHO)$_n$ **2a** and (i-Pr)$_2$NH **3a** in dioxane in the presence of CuBr (5 mol.%) did not afford

Table 7 Palladium-catalyzed C–H olefination

Entry	R^1	R^2	Indole	Product	Yield (%)[a]
1	CF$_3$	H	**7s**	**9b**	64
2	CO$_2$Me	H	**7t**	**9c**	54
3	CH$_3$	H	**7u**	**9d**	62
4	H	CF$_3$	**7v**	**9e**	62
5	H	CO$_2$Me	**7w**	**9f**	77

Reactions were carried out with 2-(aminomethyl)indole **7**, Pd(OAc)$_2$ (10 mol.%), PPh$_3$ (20 mol.%), and CsOAc (2 equiv) in DMA (2 mL) at 100 °C for 0.5 h
[a] Yields of isolated products

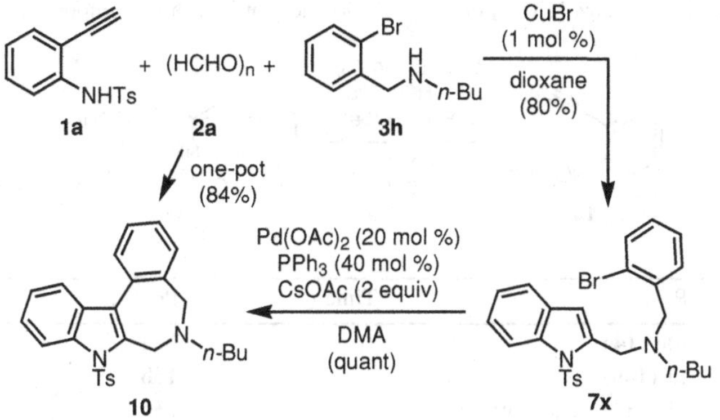

Scheme 3 Palladium-catalyzed C–H arylation and one-pot formation of polycyclic indoles from ethynylaniline

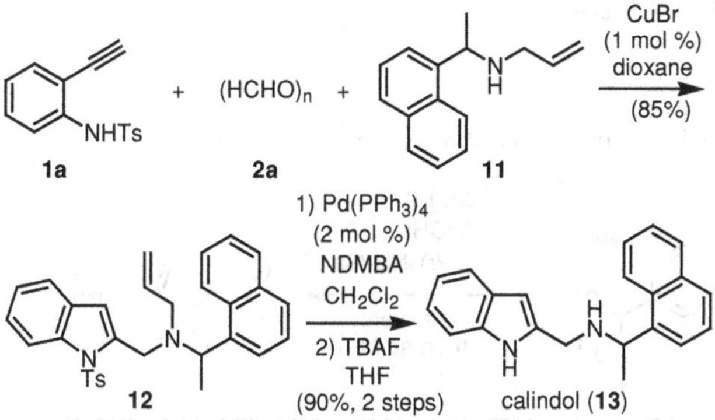

NDMBA = *N,N*'-dimethylbarbituric acid.

Scheme 4 Synthesis of calindol

the desired indene **17**, only to give the Mannich adduct in 90% yield (entry 1). A careful evaluation of the reaction conditions revealed that the use of more polar DMF as the solvent converted **16** into the desired 2-(aminomethyl)indene **17** in 39% yield. Addition of (*i*-Pr)$_2$NEt after completion of the Mannich reaction efficiently promoted the indene formation leading to **17** in 70% yield.

In conclusion, the author has developed a novel synthesis of 2-(aminomethyl)indoles through a copper-catalyzed domino three-component coupling–cyclization. This domino reaction forming two carbon–nitrogen bonds and one carbon–carbon bond is the first catalytic multi-component indole construction producing water as the only theoretical waste. The use of the chiral ligand PINAP in the reaction with alkyl aldehydes produced the corresponding indole bearing a

Table 8 Synthesis of benzo[e][1,2]thiazine-1,1-dioxide motif by three-component coupling and cyclization

Entry	R	Time (h)	Product	Yield (%)[a]
1	Me (**14a**)	16	**15a**	34
2	Et (**14b**)	22	**15b**	37
3	(4-CH$_3$)C$_6$H$_4$ (**14c**)	3.5	**15c**	90
4	Ph (**14d**)	4	**15d**	92
5	(4-MeO)C$_6$H$_4$ (**14e**)	3.5	**15e**	89
6	(4-Cl)C$_6$H$_4$ (**14f**)	3	**15f**	95

Reactions were carried out with **2a** (2.0 equiv) and **3a** (1.2 equiv) in the presence of CuBr (5 mol.%) in 1,4-dioxane (3 mL) at 100 °C
[a] Yields of isolated products

Table 9 Synthesis of 2-(aminomethyl)indene **17**

Entry	Solvent	Additive[a]	Temperature (°C)	Time (h)	Yield (%)[b]
1	Dioxane	–	80	2	0
2	DMF	–	150	5	39
3	DMF	(i-Pr)$_2$NEt	110	10	70

Reactions were carried out with **2a** (2.0 equiv) and **3a** (1.2 equiv) in solvent (2 mL) in the presence of CuBr (5 mol.%)
[a] Added after completion of the Mannich-type reaction (ca. 30 min, monitored by TLC), [b] yields of isolated products

branched substituent **7g** with moderate ee values. This reaction is synthetically useful for diversity-oriented synthesis of not only 2-(aminomethyl)indoles but also tetrahydropyridine- and benzazepine-fused indoles, using readily available reaction components. The benzo[e][1,2]thiazine and indene motif could also be constructed using a similar domino three-component coupling and cyclization strategy.

2.2 Experimental Section

2.2.1 General Methods

^{1}H NMR spectra were recorded at 400 or 500 MHz frequency, respectively. Chemical shifts are reported in δ (ppm) relative to Me$_4$Si (in CDCl$_3$) as internal standard. ^{13}C NMR spectra were referenced to the residual CHCl$_3$ signal. Melting points were measured by a hot stage melting points apparatus (uncorrected). COSY spectra (for confirmation of the NMR peak assignments) were recorded at 500 MHz frequency.

The compound **1a** [see footnote 1, 36], **S1** [see footnote 2, 37], and **S12** [see footnote 3, 38], were synthesized according to the literature.

The compounds **S7a–e**, **S9**, and **S10a, b** are commercially available.

The compounds **S7a** [39], **S7b** [40], **S7c** [41], **S7d** [42], **S9d** [43], and **S12** [44] are known.

2.2.1.1 N-(tert-Butoxycarbonyl)-2-iodo-N-tosylaniline (S2)

To a stirred solution of **S1** (0.82 g, 2.19 mmol), DMAP (54.0 mg, 0.44 mmol) in acetonitrile (9 mL) was added Boc$_2$O (0.72 g, 3.29 mmol) at rt under argon, and the reaction mixture was stirred for 0.5 h at this temperature. The reaction mixture was stirred at 80 °C for 15 h. After concentration under reduced pressure, the residue was extracted with Et$_2$O. The extract was washed successively with aqueous saturated NaHCO$_3$ and brine, and dried over MgSO$_4$. The filtrate was concentrated under reduced pressure and the residue was purified by column chromatography over alumina with hexane–EtOAc (3:1) to give **S2** (561 mg, 54%) as a colorless solid which was recrystallized from hexane–CHCl$_3$ to give pure **S2** as colorless crystals: mp 113 °C; IR (neat) cm^{-1} 1734 (C=O); ^{1}H NMR (500 MHz, CDCl$_3$) δ 1.38 (s, 9H, C(CH$_3$)$_3$), 2.46 (s, 3H, ArCH$_3$), 7.08–7.11 (m, 1H, Ar), 7.34–7.37 (m, 3H, Ar), 7.39–7.43 (m, 1H, Ar), 7.91 (dd, J = 8.0, 1.7 Hz, 1H, Ar), 8.01 (d, J = 8.6 Hz, 2H, Ar); ^{13}C NMR (125 MHz, CDCl$_3$) δ 21.7, 27.9 (3C), 84.6, 101.1, 129.0, 129.2 (2C), 129.5 (2C), 130.3, 130.8, 136.6, 139.6, 139.9, 144.8, 149.7. Anal. Calcd for C$_{18}$H$_{20}$INO$_4$S: C, 45.68; H, 4.26; N, 2.96. Found: C, 45.71; H, 4.18; N, 2.72.

2.2.1.2 N-(*tert*-Butoxycarbonyl)-N-tosyl-2-[(trimethylsilyl)ethynyl]aniline (S3)

To a stirred suspension of **S2** (0.51 g, 1.08 mmol), PdCl$_2$(PPh$_3$)$_2$ (38.0 mg, 0.054 mmol) and CuI (10.2 mg, 0.054 mmol) in a mixed solvent of THF (5 mL) and Et$_3$N (5 mL) was added TMS-acetylene (0.18 mL, 1.30 mmol) at rt under argon, and the reaction mixture was stirred for at 80 °C 12 h. The mixture was filtered through a pad of Celite. The filtrate was concentrated under reduced pressure and the residue was purified by column chromatography over silica gel with hexane–EtOAc (10:1) to give **S3** (206 mg, 43%) as a colorless solid. Recrystallization from hexane–CHCl$_3$ gave pure **S3** as colorless crystals: mp 79–80 °C; IR (neat) cm^{-1} 2162 (C≡C), 1736 (C=O); ^{1}H NMR (500 MHz, CDCl$_3$) δ 0.05 (s, 9H, Si(CH$_3$)$_3$), 1.35 (s, 9H, C(CH$_3$)$_3$), 2.44 (s, 3H, ArCH$_3$), 7.29–7.42 (m, 5H, Ar), 7.52–7.54 (m, 1H, Ar), 7.96 (d, J = 8.6 Hz, 2H, Ar); ^{13}C NMR (125 MHz, CDCl$_3$) δ 0.22 (3C), 22.2, 28.4 (3C), 84.5, 100.1, 101.4, 124.2, 129.2, 129.6, 129.7 (2C), 130.0 (2C), 131.5, 134.1, 137.7, 138.4, 144.8, 150.7. Anal. Calcd for C$_{23}$H$_{29}$NO$_4$SSi: C, 62.27; H, 6.59; N, 3.16. Found: C, 62.28; H, 6.58; N, 3.10.

2.2.1.3 N-(*tert*-Butoxycarbonyl)-2-ethynyl-N-tosylaniline (S4)

To a solution of **S3** (140 mg, 0.32 mmol) in THF (2 mL) was added TBAF (1 M in THF, 0.34 mL, 0.33 mmol) at −78 °C and the reaction mixture was stirred for 2 min at this temperature. After quenching with aqueous saturated citric acid, the whole was extracted with Et$_2$O. The extract was washed with water, aqueous saturated NaHCO$_3$ and brine, and dried over MgSO$_4$. Usual workup followed by purification by column chromatography over silica gel with hexane–EtOAc (5:1) gave **S4** (73.4 mg, 62%) as a colorless solid, which was recrystallized from hexane–CHCl$_3$ to give pure **S4** as colorless crystals: mp 133–133 °C; IR (neat) cm^{-1} 2110 (C≡C), 1732 (C=O); ^{1}H NMR (500 MHz, CDCl$_3$) δ 1.34 (s, 9H, C(CH$_3$)$_3$), 2.45 (s, 3H, ArCH$_3$), 2.91 (s, 1H, CH), 7.31 (d, J = 8.6 Hz, 2H, Ar), 7.35–7.45 (m, 3H, Ar), 7.53–7.55 (m, 1H, Ar), 7.95–7.97 (m, 2H, Ar); ^{13}C NMR (125 MHz, CDCl$_3$) δ 21.7, 27.8 (3C), 79.9, 82.1, 84.3, 122.7, 128.8, 129.0 (2C), 129.4 (2C), 129.5, 130.9, 133.3, 136.6, 138.5, 144.5, 150.2. Anal. Calcd for C$_{20}$H$_{21}$NO$_4$S: C, 64.67; H, 5.70; N, 3.77. Found: C, 64.40; H, 5.61; N, 3.72.

2.2.1.4 N-(*tert*-Butoxycarbonyl)-2-[3-(piperidin-1-yl)propy-1-nyl]-N-tosylaniline (S5)

To a stirred solution of **S4** (200 mg, 0.54 mmol), (HCHO)$_n$ (32.4 mg, 1.08 mmol), and CuBr (3.9 mg, 0.027 mmol) in dioxane (5 mL) was added piperidine (64.0 μL, 0.65 mmol) at rt under argon. The reaction mixture was stirred at 80 °C for 10 min. Concentration under reduced pressure followed by purification by column chromatography over silica gel with hexane–EtOAc (3:1) gave **S5**

(253 mg, quant) as a pale yellow solid, which was recrystallized from hexane–CHCl$_3$ to give pure **S5** as pale yellow oil: IR (neat) cm^{-1} 2233 (C≡C), 1733 (C=O); ^{1}H NMR (500 MHz, CDCl$_3$) δ 1.34 (s, 9H, C(CH$_3$)$_3$), 1.40–1.44 (m, 2H, CH$_2$), 1.57–1.61 (m, 4H, 2 × CH$_2$), 2.41–2.47 (s, 7H, 2 × CH$_2$ and ArCH$_3$), 3.15 (s, 2H, CH$_2$), 7.28–7.37 (m, 5H, Ar), 7.51 (d, $J = 7.4$ Hz, 1H, Ar), 7.97 (d, $J = 8.6$ Hz, 2H, Ar); ^{13}C NMR (125 MHz, CDCl$_3$) δ 21.5, 23.6, 25.8 (2C), 27.7 (3C), 48.2, 53.2 (2C), 81.4, 83.9, 90.2, 123.7, 128.5, 128.7, 128.9 (2C), 129.2 (2C), 130.6, 132.9, 136.9, 137.7, 144.1, 150.2; MS (FAB) m/z: 469 (MH$^+$, 100); HRMS (FAB) calcd for C$_{26}$H$_{33}$N$_2$O$_4$S (MH$^+$), 469.2161; found, 469.2161.

2.2.1.5 2-[3-(Piperidin-1-yl)prop-1-ynyl]-*N*-tosylaniline (8)

To a stirred mixture of **S5** (150 mg, 0.32 mmol) and water (75 μL) in chloroform (1.5 mL) was added TFA (1.5 mL) at 0 °C. The reaction mixture was stirred for 2.5 h at this temperature. After concentration under reduced pressure, the residue was quenched with aqueous saturated NaHCO$_3$. The whole was extracted with CH$_2$Cl$_2$, and the extract was dried over MgSO$_4$. Usual workup followed by purification by column chromatography over alumina with hexane–EtOAc (7:1) then CHCl$_3$–CH$_3$OH (10:1) gave **8** (53.8 mg, 45%) as a colorless solid which was recrystallized from hexane–CHCl$_3$ to give pure **8** as colorless crystals: mp 111 °C; IR (neat) cm^{-1} 3266 (NH), 2256 (C≡C); ^{1}H NMR (500 MHz, CDCl$_3$) δ 1.45–1.49 (m, 2H, CH$_2$), 1.64–1.68 (m, 4H, 2 × CH$_2$), 2.37 (s, 3H, ArCH$_3$), 2.51–2.55 (s, 4H, 2 × CH$_2$), 3.50 (s, 2H, CH$_2$), 6.98–7.01 (m, 1H, Ar), 7.20–7.31 (m, 5H, Ar), 7.58 (d, $J = 8.0$ Hz, 1H, Ar), 7.67 (d, $J = 8.6$ Hz, 1H, Ar); ^{13}C NMR (125 MHz, CDCl$_3$) δ 21.6, 23.8, 25.9 (2C), 48.5, 53.5 (2C), 79.8, 92.4, 113.9, 119.2, 124.1, 127.2 (2C), 129.4, 129.6 (2C), 132.2, 136.2, 137.8, 144.0. Anal. Calcd for C$_{21}$H$_{24}$N$_2$O$_2$S: C, 68.45; H, 6.56; N, 7.60. Found: C, 68.25; H, 6.56; N, 7.50.

2.2.1.6 Synthesis of 2-[(Piperidin-1-yl)methyl]-1-tosylindole 7b from 8

To a stirred solution of **8** (25.0 mg, 0.068 mmol) in dioxane (1 mL) was added CuBr (0.5 mg, 0.0034 mmol) at rt under argon. The reaction mixture was stirred at 80 °C for 50 min. Concentration under reduced pressure followed by purification by column chromatography over silica gel with hexane–EtOAc (5:1) gave **7b** (25.0 mg, quant) as a colorless solid: mp 99 °C; ^{1}H NMR (400 MHz, CDCl$_3$) δ 1.43–1.47 (m, 2H, CH$_2$), 1.51–1.56 (m, 4H, 2 × CH$_2$), 2.33 (s, 3H, CH$_3$), 2.46–2.54 (m, 4H, 2 × CH$_2$), 3.84 (s, 2H, ArCH$_2$), 6.54 (s, 1H, 3-H), 7.17–7.27 (m, 4H, Ar), 7.43–7.45 (m, 1H, Ar), 8.03 (d, $J = 8.0$ Hz, 2H, Ar), 8.07 (d, $J = 8.0$ Hz, 1H, Ar); ^{13}C NMR (100 MHz, CDCl$_3$) δ 21.5, 24.3, 25.9 (2C), 54.6 (2C), 56.2, 111.2, 114.5, 120.4, 123.2, 124.0, 127.2 (2C), 129.0, 129.4 (2C), 136.5, 137.1, 138.4, 144.4; MS (FAB) m/z (%): 369 (MH$^+$, 100), 284 (20); HRMS (FAB) calcd for C$_{21}$H$_{25}$N$_2$O$_2$S (MH$^+$): 369.1637; found: 369.1632.

S6a (R = 4-CF$_3$, X = I)
S6b (R = 4-CO$_2$CH$_3$, X = I)
S6c (R = 4-CH$_3$, X = I)
S6d (R = 5-CF$_3$, X = Br)
S6e (R = 5-CO$_2$CH$_3$, X = I)

S7a (R = 4-CF$_3$)
S7b (R = 4-CO$_2$CH$_3$)
S7c (R = 4-CH$_3$)
S7d (R = 5-CF$_3$)
S7e (R = 5-CO$_2$CH$_3$)

1b (R = 4-CF$_3$)
1c (R = 4-CO$_2$CH$_3$)
1d (R = 4-CH$_3$)
1e (R = 5-CF$_3$)
1f (R = 5-CO$_2$CH$_3$)

2.2.1.7 2-Ethynyl-N-(p-toluenesulfonyl)-4-(trifluoromethyl)aniline (1b)

To a stirred suspension of **S6a** (1.50 g, 5.23 mmol), PdCl$_2$(PPh$_3$)$_2$ (91.7 mg, 0.13 mmol) and CuI (24.9 mg, 0.13 mmol) in THF (1 mL) and Et$_3$N (20 mL) was added TMS-acetylene (0.86 mL, 6.27 mmol) at rt under argon, and the reaction mixture was stirred for 0.5 h at this temperature. The mixture was filtered through a pad of Celite. The filtrate was concentrated under reduced pressure and the residue was purified by column chromatography over silica gel with hexane–EtOAc (20:1) to give the known compound **S7a** (1.30 g, 96%).

To a stirred solution of **S7a** (1.50 g, 5.82 mmol) in pyridine (10 mL) was added TsCl (1.66 g, 8.73 mmol) at 0 °C under argon and the reaction mixture was stirred overnight at rt. After concentration under reduced pressure, the residue was extracted with EtOAc. The extract was washed successively with 3 N HCl and brine, and dried over MgSO$_4$. Usual workup followed by purification over silica gel with hexane–EtOAc (20:1) gave crude tosylate as a pale yellow solid, which was used in the next step without further purification. To a stirred mixture of the tosylate in THF (10 mL) and water (0.5 mL) was treated with TBAF (1 M in THF, 5.2 mL, 5.20 mmol) at 0 °C for 5 min. The reaction mixture was quenched with aqueous saturated citric acid, and the whole was extracted with EtOAc. The extract was washed successively with H$_2$O, aqueous saturated NaHCO$_3$, and brine, and dried over MgSO$_4$. Concentration under reduced pressure followed by purification through a pad of silica gel with hexane–EtOAc (5:1) gave **1b** (1.76 g, 89%) as a colorless solid, which was recrystallized from n-hexane–EtOAc to give pure **1b** as colorless crystals: mp 99 °C; IR (neat) cm^{-1} 3295 (NH), 2112 (C≡C); ^{1}H NMR (500 MHz, CDCl$_3$) δ 2.39 (s, 3H, CH$_3$), 3.51 (s, 1H, C≡CH), 7.27 (d, J = 8.0 Hz, 2H, Ar), 7.45 (br s, 1H, NH), 7.51 (dd, J = 8.6, 2.3 Hz, 1H, Ar), 7.61 (d, J = 2.3 Hz, 1H, Ar), 7.67 (d, J = 8.6 Hz, 1H, Ar), 7.73–7.75 (m, 2H, Ar); ^{13}C NMR (125 MHz, CDCl$_3$) δ 21.6, 77.3, 86.0, 112.1, 117.9, 123.4 (q, J = 272.3 Hz), 126.0 (q, J = 33.6 Hz), 127.0 (q, J = 3.6 Hz), 127.3 (2C), 129.7 (q, J = 3.6 Hz), 130.0 (2C), 135.7, 141.4, 144.7. Anal. Calcd for C$_{16}$H$_{12}$F$_3$NO$_2$S: C, 56.63; H, 3.56; N, 4.13. Found C, 56.88; H, 3.54; N, 4.14.

2.2.1.8 2-Ethynyl-4-(methoxycarbonyl)-*N*-(*p*-toluenesulfonyl)aniline (1c)

By a procedure identical to that described for of 2-(trimethylsilylethynyl)aniline **S7a**, 2-iodoaniline **S6b** (1.00 g, 3.61 mmol) was converted into the known compound **S7b** (2.80 g, 77%).

By a procedure similar to that described for of 2-ethynylaniline **1b**, **S7b** (1.64 g, 6.63 mmol) was converted into 2-ethynylaniline **1c** (1.92 g, 88%) as colorless crystals: mp 120 °C; IR (neat) cm^{-1} 3299 (NH), 2104 (C≡C), 1717 (C=O); ^{1}H NMR (500 MHz, CDCl$_3$) δ 2.38 (s, 3H, ArCH$_3$), 3.49 (s, 1H, C≡CH), 3.87 (s, 3H, OMe), 7.25 (d, J = 8.0 Hz, 2H, Ar), 7.52 (br s, 1H, NH), 7.62 (d, J = 8.8 Hz, 1H, Ar), 7.73–7.76 (m, 2H, Ar), 7.93 (dd, J = 8.8, 2.0 Hz, 1H, Ar), 8.04 (d, J = 2.0 Hz, 1H, Ar); ^{13}C NMR (125 MHz, CDCl$_3$) δ 21.5, 52.2, 77.6, 85.4, 111.6, 117.2, 125.5, 127.3 (2C), 129.9 (2C), 131.4, 134.1, 135.6, 142.2, 144.6, 165.6. Anal. Calcd for C$_{17}$H$_{15}$NO$_4$S: C, 61.99; H, 4.59; N, 4.25. Found C, 62.09; H, 4.61; N, 4.31.

2.2.1.9 2-Ethynyl-4-methyl-*N*-(*p*-toluenesulfonyl)aniline (1d)

By a procedure identical to that described for 2-(trimethylsilylethynyl)aniline **S7a**, 2-iodoaniline **S6c** (2.03 g, 3.61 mmol) was converted into the known compound **S7c** (1.77 g, quant).

By a procedure identical to that described for 2-ethynylaniline **1b**, **S7c** (0.93 g, 4.57 mmol) was converted into 2-ethynylaniline **1d** (1.17 g, 90%) as colorless crystals: mp 104 °C; IR (neat) cm^{-1} 3284 (NH), 2109 (C≡C); ^{1}H NMR (500 MHz, CDCl$_3$) δ 2.23 (s, 3H, CH$_3$), 2.36 (s, 3H, CH$_3$), 3.29 (s, 1H, C≡CH), 7.09–7.13 (m, 3H, Ar), 7.20 (d, J = 8.6 Hz, 2H, Ar), 7.48 (d, J = 8.6 Hz, 1H, Ar), 7.66 (d, J = 8.0 Hz, 2H, Ar); ^{13}C NMR (125 MHz, CDCl$_3$) δ 20.5, 21.6, 78.8, 83.8, 112.9, 119.9, 127.4 (2C), 129.6 (2C), 131.0, 132.8, 134.2, 135.9, 136.0, 144.0. Anal. Calcd for C$_{16}$H$_{15}$NO$_2$S: C, 67.34; H, 5.30; N, 4.91. Found C, 67.42; H, 5.18; N, 4.91.

2.2.1.10 2-Ethynyl-*N*-(*p*-toluenesulfonyl)-5-(trifluoromethyl)aniline (1e)

By a procedure identical to that described for the 2-(trimethylsilylethynyl)aniline **S7a**, 2-bromoaniline **S6d** (2.09 g, 8.69 mmol) was converted into the known compound **S7d** (1.70 g, 76%) by the reaction under reflux for 16 h.

By a procedure similar to that described for 2-ethynylaniline **1b**, **S7d** (2.22 g, 8.63 mmol) was converted into 2-ethynylaniline **1e** (1.69 g, 58%) as colorless crystals: mp 162 °C; IR (neat) cm^{-1} 3266 (NH), 2111 (C≡C); ^{1}H NMR (500 MHz, CDCl$_3$) δ 2.38 (s, 3H, CH$_3$), 3.51 (s, 1H, C≡CH), 7.24–7.26 (m, 3H, Ar), 7.35 (br s, 1H, NH), 7.45 (d, J = 8.0 Hz, 1H, Ar), 7.70–7.72 (m, 2H, Ar), 7.86 (s, 1H, Ar); ^{13}C NMR (125 MHz, CDCl$_3$) δ 21.6, 77.5, 86.6, 115.6 (m, 2C), 120.5 (q, J = 3.6 Hz), 123.3 (q, J = 272.3 Hz), 127.4 (2C), 129.9 (2C), 132.1 (q,

$J = 33.6$ Hz), 133.0, 135.5, 139.1, 144.7. Anal. Calcd for $C_{16}H_{12}F_3NO_2S$: C, 56.63; H, 3.56; N, 4.13. Found C, 56.77; H, 3.74; N, 4.12.

2.2.1.11 2-Ethynyl-5-(methoxycarbonyl)-N-(p-toluenesulfonyl)aniline (1f)

By a procedure identical to that described for 2-(trimethylsilylethynyl)aniline **S7a**, 2-iodoaniline **S6e** (3.00 g, 10.8 mmol) was converted into **S7e** (2.40 g, 90%) as colorless crystals.

Compound **S7e**: mp 64 °C; IR (neat) cm^{-1} 3480, 3378 (NH$_2$), 2145 (C≡C), 1713 (C=O); ^{1}H NMR (500 MHz, CDCl$_3$) δ 0.27 (s, 9H, 3 × CH$_3$), 3.88 (s, 3H, OMe), 4.34 (s, 2H, NH$_2$), 7.30–7.36 (m, 3H, Ar); ^{13}C NMR (125 MHz, CDCl$_3$) δ 0.00 (3C), 52.1, 100.9, 102.7, 112.0, 114.9, 118.6, 131.0, 132.2, 148.1, 166.8. Anal. Calcd for $C_{13}H_{17}NO_2Si$: C, 63.12; H, 6.93; N, 5.66. found C, 63.12; H, 6.93; N, 5.66.

By a identical similar to that described for 2-ethynylaniline **1b**, **S7e** (2.40 g, 9.66 mmol) was converted into 2-ethynylaniline **1f** (2.46 g, 77%) as colorless crystals: mp 160 °C; IR (neat) cm^{-1} 3268 (NH), 2105 (C≡C), 1720 (C=O); ^{1}H NMR (500 MHz, CDCl$_3$) δ 2.37 (s, 3H, ArCH$_3$), 3.51 (s, 1H, C≡CH), 3.92 (s, 3H, OMe), 7.23 (d, $J = 8.6$ Hz, 2H, Ar), 7.27 (br s, 1H, NH), 7.40 (d, $J = 8.0$ Hz, 1H, Ar), 7.68 (dd, $J = 8.0$, 1.7 Hz, 1H, Ar), 7.72 (d, $J = 8.6$ Hz, 2H, Ar), 8.23 (d, $J = 1.7$ Hz, 1H, Ar); ^{13}C NMR (125 MHz, CDCl$_3$) δ 21.6, 52.5, 78.0, 86.8, 116.7, 119.9, 125.0, 127.5 (2C), 129.8 (2C), 131.7, 132.5, 135.8, 138.7, 144.4, 165.8. Anal. Calcd for $C_{17}H_{15}NO_4S$: C, 61.99; H, 4.59; N, 4.25. Found C, 62.25; H, 4.56; N, 4.30.

2.2.2 General Procedure for Synthesis of 2-(Aminomethyl)indole

2.2.2.1 Synthesis of 2-[(N,N-Diisopropylamino)methyl]-1-tosylindole (7a)

To a stirred mixture of 2-ethynylaniline **1a** (50.0 mg, 0.18 mmol), (HCHO)$_n$ (11.1 mg, 0.37 mmol), and CuBr (0.3 mg, 0.0018 mmol) in dioxane (3.0 mL) was added diisopropylamine **3a** (28.6 μL, 0.20 mmol) at rt under argon, and the reaction mixture was stirred at 80 °C for 15 min. Concentration under reduced pressure followed by purification by column chromatography over silica gel with hexane–EtOAc (10:1) afforded the indole **7a** (57.3 mg, 81%) as a colorless solid: mp 105 °C; ^{1}H NMR (400 MHz, CDCl$_3$) δ 0.98 (d, $J = 6.6$ Hz, 12H, 4 × CHCH$_3$), 2.33 (s, 3H, ArCH$_3$), 3.01–3.11 (m, 2H, 2 × CH), 3.92 (d, $J = 1.5$ Hz, 2H, CH$_2$), 6.79 (s, 1H, 3-H), 7.18–7.25 (m, 4H, Ar), 7.41–7.43 (m, 1H, Ar), 7.64–7.67 (m, 2H, Ar), 8.16 (d, $J = 8.0$ Hz, 1H); ^{13}C NMR (100 MHz, CDCl$_3$) δ 20.8 (4C), 21.5, 44.4, 49.3 (2C), 110.1, 114.4, 120.2, 123.3, 123.5, 126.3 (2C), 129.8 (2C), 129.9, 136.4, 137.8, 144.4, 144.6; MS (FAB) m/z (%): 385

(MH$^+$, 100), 284 (75); HRMS (FAB) calcd for $C_{22}H_{29}N_2O_2S$ (MH$^+$): 385.1950; found: 385.1953.

2.2.2.2 2-[(Piperidin-1-yl)methyl]-1-tosylindole (7b) from 1a

By a procedure similar to that described for indole **7a**, **1a** (50.0 mg, 0.18 mmol) was converted into **7b** (59.2 mg, 87%) using piperidine **3b** (20.0 μL, 0.20 mmol).

2.2.2.3 2-[(Pyrrolidin-1-yl)methyl]-1-tosylindole (7c)

By a procedure similar to that described for indole **7a**, **1a** (50.0 mg, 0.18 mmol) was converted into **7c** (59.2 mg, 89%) as a colorless solid using pyrrolidine **3c** (16.8 μL, 0.20 mmol): mp 114 °C; ^{1}H NMR (400 MHz, CDCl$_3$) δ 1.71–1.77 (m, 4H, 2 × CH$_2$), 2.32 (s, 3H, CH$_3$), 2.56–2.60 (m, 4H, 2 × CH$_2$), 4.04 (s, 2H, ArCH$_2$), 6.58 (d, J = 0.5 Hz, 1H, 3-H), 7.15–7.29 (m, 4H, Ar), 7.43–7.45 (m, 1H, Ar), 7.87–7.90 (m, 2H, Ar), 8.12 (dd, J = 8.3, 1.0 Hz, 1H, Ar); ^{13}C NMR (100 MHz, CDCl$_3$) δ 21.5, 23.6 (2C), 53.1, 54.0 (2C), 110.4, 114.6, 120.4, 123.2, 124.1, 126.9 (2C), 129.2, 129.4 (2C), 136.5, 137.1, 139.3, 144.4; MS (FAB) m/z (%): 355 (MH$^+$, 100), 284 (20); HRMS (FAB) calcd for $C_{20}H_{23}N_2O_2S$ (MH$^+$): 355.1480; found: 355.1485.

2.2.2.4 2-[(N,N-Diethylamino)methyl]-1-tosylindole (7d)

By a procedure similar to that described for of indole **7a**, **1a** (50.0 mg, 0.18 mmol) was converted into **7d** (58.2 mg, 89%) as a colorless solid using diethylamine **3d** (38.1 μL, 0.37 mmol): mp 51 °C; ^{1}H NMR (400 MHz, CDCl$_3$) δ 0.99 (t, J = 7.1 Hz, 6H, 2 × CH$_2$CH_3), 2.33 (s, 3H, ArCH$_3$), 2.60 (q, J = 7.1 Hz, 4H, 2 × CH$_2$CH$_3$), 3.94 (s, 2H, ArCH$_2$), 6.62 (s, 1H, 3-H), 7.17–7.27 (m, 4H, Ar), 7.44 (d, J = 7.1 Hz, 1H, Ar), 7.85 (d, J = 8.3 Hz, 2H, Ar), 8.12 (d, J = 8.3 Hz, 1H, Ar); ^{13}C NMR (100 MHz, CDCl$_3$) δ 11.2 (2C), 21.5, 46.7 (2C), 51.5, 111.0, 114.6, 120.4, 123.3, 124.0, 126.8 (2C), 129.3, 129.5 (2C), 136.4, 137.3, 139.7, 144.5; MS (FAB) m/z (%): 357 (MH$^+$, 100), 284 (60); HRMS (FAB) calcd for $C_{20}H_{25}N_2O_2S$ (MH$^+$): 357.1637; found: 357.1633.

2.2.2.5 2-[(N,N-Diallylamino)methyl]-1-tosylindole (7e)

By a procedure similar to that described for indole **7a**, **1a** (50.0 mg, 0.18 mmol) was converted into **7e** (54.8 mg, 78%) as a colorless solid using diallylamine **3e** (25.0 μL, 0.20 mmol) (30 min): mp 42 °C; ^{1}H NMR (500 MHz, CDCl$_3$) δ 2.33 (s, 3H, CH$_3$), 3.18–3.23 (m, 4H, 2 × NCH$_2$), 4.02 (s, 2H, ArCH$_2$), 5.14–5.22 (m, 4H,

$2 \times$ CH=CH_2), 5.82–5.89 (m, 2H, $2 \times$ CH=CH$_2$), 6.71 (s, 1H, 3-H), 7.17 (d, $J = 8.6$ Hz, 2H, Ar), 7.19–7.22 (m, 1H, Ar), 7.25–7.28 (m, 1H, Ar), 7.45 (d, $J = 7.4$ Hz, 1H, Ar), 7.75 (d, $J = 8.6$ Hz, 2H, Ar), 8.13 (d, $J = 8.0$ Hz, 1H, Ar); ^{13}C NMR (125 MHz, CDCl$_3$) δ 21.5, 51.2, 56.5 (2C), 110.7, 114.6, 117.8 (2C), 120.5, 123.4, 124.1, 126.7 (2C), 129.4, 129.6 (2C), 135.1 (2C), 136.2, 137.4, 139.8, 144.6; MS (FAB) m/z (%): 381 (MH$^+$, 100), 284 (75); HRMS (FAB) calcd for C$_{22}$H$_{25}$N$_2$O$_2$S (MH$^+$): 381.1637; found: 381.1640.

2.2.2.6 2-[(*N*,*N*-Dibenzylamino)methyl]-1-tosylindole (7f)

By a procedure similar to that described for indole **7a**, **1a** (50.0 mg, 0.18 mmol) was converted into **7f** (69.2 mg, 78%) as a colorless solid by treatment with dibenzylamine **3f** (39 µL, 0.20 mmol) for 2 h: mp 118 °C; ^{1}H NMR (400 MHz, CDCl$_3$) δ 2.29 (s, 3H, CH$_3$), 3.73 (s, 4H, $2 \times$ CH$_2$), 4.03 (s, 2H, CH$_2$), 6.93 (s, 1H, 3-H), 7.02 (d, $J = 8.3$ Hz, 2H, Ar), 7.18–7.46 (m, 15H, Ar), 8.12 (d, $J = 8.3$ Hz, 1H, Ar); ^{13}C NMR (100 MHz, CDCl$_3$) δ 21.5, 52.0, 58.5 (2C), 109.8, 114.7, 120.4, 123.6, 123.9, 126.2 (2C), 126.9 (2C), 128.3 (4C), 128.4 (4C), 129.7 (2C), 129.9, 135.6, 137.4, 139.2 (2C), 140.1, 144.5; MS (FAB) m/z (%): 481 (MH$^+$, 100), 284 (40); HRMS (FAB) calcd for C$_{30}$H$_{29}$N$_2$O$_2$S (MH$^+$): 481.1950; found: 481.1942.

2.2.2.7 2-[1-(Piperidin-1-yl)butyl]-1-tosylindole (7g)

By a procedure similar to that described for indole **7b**, **1a** (50.0 mg, 0.18 mmol) was converted into **7g** (75.7 mg, quant) as a colorless solid using butanal **2b** (33.2 µL, 0.37 mmol): mp 109 °C; ^{1}H NMR (400 MHz, CDCl$_3$) δ 0.91 (t, $J = 7.3$ Hz, 3H, CH$_2$CH_3), 1.21–1.54 (m, 8H, $4 \times$ CH$_2$), 1.62–1.71 (m, 1H, CHH), 1.80–1.89 (m, 1H, CHH), 2.31 (s, 3H, ArCH$_3$), 2.47–2.59 (m, 4H, $2 \times$ NCH$_2$), 4.70 (dd, $J = 9.8$, 4.6 Hz, 1H, NCH), 6.51 (s, 1H, 3-H), 7.13–7.26 (m, 4H, Ar), 7.44–7.46 (m, 1H, Ar), 7.92 (d, $J = 8.3$ Hz, 2H, Ar), 8.08 (d, $J = 8.3$ Hz, 1H, Ar); ^{13}C NMR (100 MHz, CDCl$_3$) δ 14.2, 20.1, 21.5, 24.7, 26.2 (2C), 30.3, 49.5 (2C), 59.8, 109.4, 115.2, 120.4, 123.3, 124.0, 127.0 (2C), 129.1, 129.3 (2C), 136.5, 137.1, 141.8, 144.3; MS (FAB) m/z (%): 411 (MH$^+$, 90), 367 (100), 326 (50); HRMS (FAB) calcd for C$_{24}$H$_{31}$N$_2$O$_2$S (MH$^+$): 411.2106; found: 411.2115.

2.2.2.8 2-[2-Methyl-1-(piperidin-1-yl)propyl]-1-tosylindole (7h)

By a procedure similar to that described for indole **7b**, **1a** (50.0 mg, 0.18 mmol) was converted into **7h** (58.3 mg, 77%) as a colorless solid by treatment with *i*-butyraldehyde **2c** (33.6 µL, 0.37 mmol) under reflux for 3 h: mp 98 °C; ^{1}H NMR (400 MHz, CDCl$_3$) δ 0.75 (d, $J = 6.6$ Hz, 3H, CCH$_3$), 1.12 (d, $J = 6.6$ Hz, 3H, CCH$_3$), 1.23–1.29 (m, 2H, CH$_2$), 1.46–1.52 (m, 4H, $2 \times$ CH$_2$), 2.08–2.19 (m, 1H, CH), 2.29 (s, 3H, ArCH$_3$), 2.32–2.38 (m, 4H, $2 \times$ CH$_2$), 4.36 (d, $J = 10.7$ Hz, 1H,

NCH), 6.40 (s, 1H, 3-H), 7.11 (d, $J = 8.0$ Hz, 2H, Ar), 7.20–7.29 (m, 2H, Ar), 7.45 (d, $J = 7.8$ Hz, 1H, Ar), 7.59 (d, $J = 8.0$ Hz, 2H, Ar), 8.16 (d, $J = 7.8$ Hz, 1H, Ar); ^{13}C NMR (100 MHz, CDCl$_3$) δ 20.7, 21.1, 21.5, 24.7, 26.7 (2C), 29.7, 49.9 (2C), 66.5, 109.9, 115.8, 120.4, 123.6, 123.9, 126.7 (2C), 129.4, (2C), 129.6, 136.2, 137.3, 140.3, 144.5; MS (FAB) m/z (%): 411 (MH$^+$, 70), 367 (100), 326 (35); HRMS (FAB) calcd for C$_{24}$H$_{31}$N$_2$O$_2$S (MH$^+$): 411.2106; found: 411.2112.

2.2.2.9 2-[Phenyl(piperidin-1-yl)methyl]-1-tosylindole (7i)

By a procedure similar to that described for indole **7b**, **1a** (50.0 mg, 0.18 mmol) was converted into **7i** (57.1 mg, 70%) as an yellow oil by treatment with benz-aldehyde **2d** (37.6 μL, 0.37 mmol) under reflux for 10 h: ^{1}H NMR (400 MHz, CDCl$_3$) δ 1.38–1.57 (m, 6H, 3 × CH$_2$), 2.23–2.30 (m, 5H, ArCH$_3$ and 2 × CHH), 2.41–2.46 (m, 2H, 2 × CHH), 5.34 (s, 1H, NCH), 6.95 (s, 1H, 3-H), 7.00 (d, $J = 8.0$ Hz, 2H, Ar), 7.18–7.30 (m, 7H, Ar), 7.37–7.39 (m, 2H, Ar), 7.46–7.48 (m, 1H, Ar), 8.08 (d, $J = 8.0$ Hz, 1H, Ar); ^{13}C NMR (100 MHz, CDCl$_3$) δ 21.4, 24.7, 26.4 (2C), 53.1 (2C), 67.5, 110.6, 115.2, 120.6, 123.5, 124.0, 126.5 (2C), 127.2, 128.0 (2C), 129.4 (2C), 129.6 (2C), 129.8, 136.0, 137.2, 140.1, 143.9, 144.4; MS (FAB) m/z (%): 445 (MH$^+$, 90), 360 (100); HRMS (FAB) calcd for C$_{27}$H$_{29}$N$_2$O$_2$S (MH$^+$): 445.1950; found: 445.1956.

2.2.2.10 2-{[4-(Methoxycarbonyl)phenyl](piperidin-1-yl)methyl}-1-tosylin-dole (7j)

By a procedure similar to that described for indole **7b**, **1a** (50.0 mg, 0.18 mmol) was converted into **7j** (70.2 mg, 76%) as a colorless solid by treatment with 4-methoxycarbonylbenzaldehyde **2e** (60.5 mg, 0.37 mmol) under reflux for 3 h: mp 167 °C; ^{1}H NMR (400 MHz, CDCl$_3$) δ 1.38–1.56 (m, 6H, 3 × CH$_2$), 2.25–2.31 (m, 5H, ArCH$_3$ and 2 × CHH), 2.39–2.44 (m, 2H, 2 × CHH), 3.90 (s, 3H, OMe), 5.42 (s, 1H, NCH), 6.90 (s, 1H, 3-H), 7.03 (d, $J = 8.0$ Hz, 2H, Ar), 7.20–7.28 (m, 2H, Ar), 7.35 (d, $J = 8.0$ Hz, 2H, Ar), 7.43 (d, $J = 8.0$ Hz, 2H, Ar), 7.47–7.50 (m, 1H, Ar), 7.89 (d, $J = 8.0$ Hz, 2H, Ar), 8.13 (d, $J = 8.0$ Hz, 1H, Ar); ^{13}C NMR (100 MHz, CDCl$_3$) δ 21.4, 24.6, 26.4 (2C), 52.0, 53.0 (2C), 67.0, 111.2, 115.3, 120.7, 123.7, 124.3, 126.3 (2C), 129.0, 129.31 (2C), 129.34 (2C), 129.5 (2C), 129.6, 136.0, 137.5, 142.7, 144.6, 145.6, 166.9; MS (FAB) m/z (%): 503 (MH$^+$, 55), 418 (100); HRMS (FAB) calcd for C$_{29}$H$_{31}$N$_2$O$_4$S (MH$^+$): 503.2005; found: 503.2008.

2.2.2.11 2-[(Piperidin-1-yl)(p-tolyl)methyl]-1-tosylindole (7k)

By a procedure similar to that described for indole **7b**, **1a** (50.0 mg, 0.18 mmol) was converted into **7k** (68.3 mg, 85%) as an yellow oil by treatment with

4-methylbenzaldehyde **2f** (43.6 μL, 0.37 mmol) under reflux for 3 h: ^{1}H NMR (400 MHz, CDCl$_3$) δ 1.37–1.56 (m, 6H, 3 × CH$_2$), 2.24–2.34 (m, 8H, 2 × ArCH$_3$ and 2 × CHH), 2.39–2.46 (m, 2H, 2 × CHH), 5.28 (s, 1H, NCH), 6.94 (s, 1H, 3-H), 6.99 (d, J = 8.0 Hz, 2H, Ar), 7.04 (d, J = 7.8 Hz, 2H, Ar), 7.18–7.31 (m, 6H, Ar), 7.46–7.48 (m, 1H, Ar), 8.08 (d, J = 8.0 Hz, 1H, Ar); ^{13}C NMR (100 MHz, CDCl$_3$) δ 21.1, 21.4, 24.7, 26.4 (2C), 53.2 (2C), 67.2, 110.4, 115.2, 120.6, 123.5, 123.9, 126.5 (2C), 128.7 (2C), 129.3 (2C), 129.5 (2C), 129.8, 136.1, 136.8, 137.0, 137.3, 144.1, 144.3; MS (FAB) m/z (%): 459 (MH$^+$, 50), 374 (100); HRMS (FAB) calcd for C$_{28}$H$_{31}$N$_2$O$_2$S (MH$^+$): 459.2106; found: 459.2114.

2.2.2.12 2-[(2-Bromophenyl)(piperidin-1-yl)methyl]-1-tosylindole (7l)

By a procedure similar to that described for indole **7b**, **1a** (50.0 mg, 0.18 mmol) was converted into **7l** (62.7 mg, 65%) as a colorless solid by treatment with 2-bromobenzaldehyde **2f** (42.7 μL, 0.37 mmol) under reflux for 4 h: mp 190 °C; ^{1}H NMR (400 MHz, CDCl$_3$) δ 1.43–1.50 (m, 6H, 3 × CH$_2$), 2.29 (s, 3H, ArCH$_3$), 2.41–2.46 (m, 2H, 2 × CHH), 2.60–2.65 (m, 2H, 2 × CHH), 5.82 (s, 1H, NCH), 6.96 (s, 1H, 3-H), 7.04–7.14 (m, 4H, Ar), 7.20–7.29 (m, 3H, Ar), 7.45–7.51 (m, 3H, Ar), 7.58–7.61 (m, 1H, Ar), 8.11 (d, J = 8.0 Hz, 1H, Ar); ^{13}C NMR (100 MHz, CDCl$_3$) δ 21.5, 24.7, 26.9 (2C), 51.9 (2C), 65.5, 112.1, 115.0, 120.7, 123.4, 124.2, 126.5 (2C), 126.8, 127.1, 128.6, 129.2, 129.4 (2C), 131.0, 133.1, 136.3, 137.7, 139.2, 143.3, 144.4; MS (FAB) m/z (%): 525 [MH$^+$ (^{81}Br), 15], 523 [MH$^+$ (^{79}Br), 15], 440 (15), 438 (15); HRMS (FAB) calcd for C$_{27}$H$_{28}$BrN$_2$O$_2$S [MH$^+$ (^{79}Br)]: 523.1055; found: 523.1052.

2.2.2.13 Enantioselective Synthesis of 2-[1-(Piperidin-1-yl)butyl]-1-tosylindole (7g)

To a stirred suspension of CuBr (1.3 mg, 0.0092 mmol) in benzene (2 mL) was added (S)-PINAP (5.7 mg, 0.010 mmol) at rt under argon. After the reaction mixture was stirred for 0.5 h at this temperature, piperidine **3b** (20.0 μL, 0.20 mmol), butanal **2b** (33.2 μL, 0.37 mmol), and **1a** (50.0 mg, 0.18 mmol) were successively added and the reaction mixture was additionally stirred for 5 d at rt. Concentration under reduced pressure followed by purification by column chromatography over silica gel with hexane–EtOAc (5:1) gave **7g** (quant, 63% ee): [α]$_D^{23}$ −23.2 (c 1.00, CHCl$_3$).

2.2.2.14 2-[(N,N-Dibenzylamino)methyl]-1-tosyl-5-(trifluoromethyl)indole (7m)

By a procedure identical to that described for of indole **7f**, **1b** (62.5 mg, 0.18 mmol) was converted into **7m** (91.0 mg, 90%) as a colorless solid by the reaction at 80 °C for 3 h: mp 95 °C; ^{1}H NMR (500 MHz, CDCl$_3$) δ 2.31 (s, 3H,

ArCH$_3$), 3.73 (s, 4H, 2 × NCH$_2$), 4.04 (d, J = 1.1 Hz, 2H, NCH$_2$), 7.00 (s, 1H, 3-H), 7.06 (d, J = 8.0 Hz, 2H, Ar), 7.23–7.26 (m, 2H, Ar), 7.29–7.32 (m, 4H, Ar), 7.39–7.42 (m, 6H, Ar), 7.48 (dd, J = 8.6, 1.7 Hz, 1H, Ar), 7.75 (s, 1H, Ar), 8.22 (d, J = 8.6 Hz, 1H, Ar); ^{13}C NMR (125 MHz, CDCl$_3$) δ 21.6, 51.9, 58.6 (2C), 109.4, 114.8, 117.9 (q, J = 3.6 Hz), 120.6 (q, J = 3.6 Hz), 124.6 (q, J = 272.3 Hz), 125.9 (q, J = 32.4 Hz), 126.2 (2C), 127.1 (2C), 128.40 (4C), 128.42 (4C), 129.6, 129.9 (2C), 135.4, 138.89, 138.94 (2C), 142.2, 145.1; MS (FAB) m/z (%): 547 (M–H$^+$, 70), 393 (100); HRMS (FAB) calcd for C$_{31}$H$_{26}$F$_3$N$_2$O$_2$S (M–H$^+$): 547.1667; found: 547.1665.

2.2.2.15   2-[(*N,N*-Dibenzylamino)methyl]-5-(methoxycarbonyl)-1-tosylindole (7n)

By a procedure identical to that described for of indole **7f**, **1c** (60.7 mg, 0.18 mmol) was converted into **7n** (90.7 mg, 91%) as a colorless solid by the reaction at 80 °C for 3 h: mp 104 °C; ^{1}H NMR (400 MHz, CDCl$_3$) δ 2.30 (s, 3H, ArCH$_3$), 3.73 (s, 4H, 2 × CH$_2$), 3.91 (s, 3H, OMe), 4.03 (s, 2H, CH$_2$), 7.00 (s, 1H, 3-H), 7.05 (d, J = 8.3 Hz, 2H, Ar), 7.23–7.42 (m, 12H, Ar), 7.94 (dd, J = 8.8, 1.5 Hz, 1H, Ar), 8.15–8.18 (m, 2H, Ar); ^{13}C NMR (100 MHz, CDCl$_3$) δ 21.5, 51.9, 52.1, 58.6 (2C), 109.8, 114.2, 122.6, 125.2, 125.5, 126.2 (2C), 127.0 (2C), 128.36 (4C), 128.38 (4C), 129.6, 129.8 (2C), 135.4, 139.0 (2C), 140.0, 141.6, 144.9, 167.2; MS (FAB) m/z (%): 537 (M–H$^+$, 85), 383 (100); HRMS (FAB) calcd for C$_{32}$H$_{29}$N$_2$O$_4$S (M–H$^+$): 537.1848; found: 537.1859.

2.2.2.16   2-[(*N,N*-Dibenzylamino)methyl]-5-methyl-1-tosylindole (7o)

By a procedure identical to that described for indole **7f**, **1d** (52.6 mg, 0.18 mmol) was converted into **7o** (71.2 mg, 78%) as a colorless solid by the reaction at 80 °C for 5 h, then reflux, 1 h): mp 140 °C; ^{1}H NMR (500 MHz, CDCl$_3$) δ 2.28 (s, 3H, ArCH$_3$), 2.38 (s, 3H, ArCH$_3$), 3.72 (s, 4H, 2 × NCH$_2$), 4.01 (s, 2H, NCH$_2$), 6.86 (s, 1H, 3-H), 7.01 (d, J = 8.0 Hz, 2H, Ar), 7.05 (d, J = 8.6 Hz, 1H, Ar), 7.22–7.25 (m, 3H, Ar), 7.28–7.31 (m, 4H, Ar), 7.37–7.41 (m, 6H, Ar), 7.99 (d, J = 8.6 Hz, 1H, Ar); ^{13}C NMR (125 MHz, CDCl$_3$) δ 21.2, 21.5, 51.9, 58.4 (2C), 109.8, 114.4, 120.4, 125.3, 126.2 (2C), 126.9 (2C), 128.3 (4C), 128.4 (4C), 129.6 (2C), 130.2, 133.1, 135.6 (2C), 139.2 (2C), 140.1, 144.4; MS (FAB) m/z (%): 495 (MH$^+$, 100), 298 (55); HRMS (FAB) calcd for C$_{31}$H$_{31}$N$_2$O$_2$S (MH$^+$): 495.2106; found: 495.2099.

2.2.2.17   2-[(*N,N*-Dibenzylamino)methyl]-1-tosyl-6-(trifluoromethyl)indole (7p)

By a procedure identical to that described for of indole **7f**, **1e** (62.5 mg, 0.18 mmol) was converted into **7p** (61.7 mg, 61%) as a colorless solid by the reaction at 80 °C, 3 h: mp 104 °C; ^{1}H NMR (500 MHz, CDCl$_3$) δ 2.31 (s, 3H, ArCH$_3$), 3.73 (s, 4H, 2 × NCH$_2$), 4.04 (s, 2H, NCH$_2$), 6.98 (s, 1H, 3-H), 7.06 (d, J = 8.0 Hz, 2H, Ar), 7.23–7.26 (m, 2H, Ar), 7.29–7.32 (m, 4H, Ar), 7.39–7.42 (m, 6H, Ar), 7.45 (dd, J = 8.0, 1.1 Hz, 1H, Ar), 7.53 (d, J = 8.0 Hz, 1H, Ar), 8.43 (s, 1H, Ar); ^{13}C NMR (125 MHz, CDCl$_3$) δ 21.5, 51.9, 58.6 (2C), 109.1, 112.0 (q, J = 3.6 Hz), 120.3 (q, J = 3.6 Hz), 120.7, 124.7 (q, J = 272.3 Hz), 126.0 (q, J = 32.4 Hz), 126.2 (2C), 127.1 (2C), 128.38 (4C), 128.40 (4C), 129.9 (2C), 132.4, 135.3, 136.5, 138.9 (2C), 143.2, 145.1; MS (FAB) m/z (%): 547 (M–H$^+$, 65), 393 (100); HRMS (FAB) calcd for C$_{31}$H$_{26}$F$_3$N$_2$O$_2$S (M–H$^+$): 547.1667; found: 547.1672.

2.2.2.18 2-[(N,N-Dibenzylamino)methyl]-6-(methoxycarbonyl)-1-tosylindole (7q)

By a procedure identical to that described for indole **7f**, **1f** (60.7 mg, 0.18 mmol) was converted into **7q** (78.3 mg, 79%) as a colorless solid by the reaction at 80 °C for 5 h: ^{1}H NMR (400 MHz, CDCl$_3$) δ 2.30 (s, 3H, ArCH$_3$), 3.73 (s, 4H, 2 × NCH$_2$), 3.94 (s, 3H, OMe), 4.06 (s, 2H, NCH$_2$), 6.98 (s, 1H, 3-H), 7.05 (d, J = 8.5 Hz, 2H, Ar), 7.23–7.49 (m, 14H, Ar), 7.91 (dd, J = 8.3, 1.5 Hz, 1H, Ar); ^{13}C NMR (100 MHz, CDCl$_3$) δ 21.5, 52.0, 52.1, 58.6 (2C), 109.4, 116.3, 120.1, 124.9, 125.8, 126.3 (2C), 127.0 (2C), 128.4 (8C), 129.8 (2C), 133.6, 135.4, 136.9, 139.0 (2C), 143.7, 144.9, 167.4; MS (FAB) m/z (%): 539 (MH$^+$, 100); HRMS (FAB) calcd for C$_{32}$H$_{31}$N$_2$O$_4$S (MH$^+$): 539.2005; found: 539.2007.

2.2.2.19 2-{[N-(2-Bromoprop-2-en-1-yl)-N-butylamino]methyl}-1-tosylindole (7r)

By a procedure similar to that described for indole **7a**, **1a** (100 mg, 0.37 mmol) was converted into **7r** (171 mg, 98%) as an yellow oil by treatment with N-(2-bromoprop-2-enyl)-N-butylamine **3g** (77.8 mg, 0.41 mmol) at 80 °C for 3 h, then in the presence of Et$_3$N (205.5 μL, 1.47 mmol) under reflux for 1 h: ^{1}H NMR (500 MHz, CDCl$_3$) δ 0.87 (t, J = 7.4 Hz, 3H, CH$_2$CH_3), 1.25–1.33 (m, 2H, CH$_2$), 1.40–1.46 (m, 2H, CH$_2$), 2.33 (s, 3H, ArCH$_3$), 2.57–2.60 (m, 2H, NCH$_2$), 3.38 (s, 2H, NCH$_2$), 4.05 (d, J = 1.1 Hz, 2H, NCH$_2$), 5.54 (s, 1H, C=CHH), 5.89 (d, J = 1.1 Hz, 1H, C=CHH), 6.81 (s, 1H, 3-H), 7.17–7.27 (m, 4H, Ar), 7.44–7.46 (m, 1H, Ar), 7.63–7.66 (m, 2H, Ar), 8.12–8.14 (m, 1H, Ar); ^{13}C NMR (125 MHz, CDCl$_3$) δ 14.1, 20.5, 21.5, 29.4, 52.2, 53.8, 62.6, 110.3, 114.5, 117.9, 120.5, 123.5, 124.0, 126.3 (2C), 129.69, 129.74 (2C), 131.9, 136.0, 137.4, 139.9, 144.7; MS (FAB) m/z (%): 475 [M–H$^+$ (^{81}Br), 100], 473 [M–H$^+$ (^{79}Br), 90]; HRMS (FAB) calcd for C$_{23}$H$_{26}$BrN$_2$O$_2$S [M–H$^+$ (^{79}Br)]: 473.0898; found: 473.0900.

2.2.2.20 2-{[N-(2-Bromoprop-2-en-1-yl)-N-butylamino]methyl}-1-tosyl-5-tri-fluoromethylindole (7s)

By a procedure identical to that described for of indole **7r**, **1b** (125 mg, 0.37 mmol) was converted into **7s** (182.3 mg, 91%) as an yellow oil by the reaction at 80 °C for 3 h: ^{1}H NMR (500 MHz, CDCl$_3$) δ 0.88 (t, J = 7.4 Hz, 3H, CH$_2$CH_3), 1.26–1.33 (m, 2H, CH$_2$), 1.40–1.46 (m, 2H, CH$_2$), 2.36 (s, 3H, ArCH$_3$), 2.57–2.60 (m, 2H, NCH$_2$), 3.38 (s, 2H, NCH$_2$), 4.05 (d, J = 1.1 Hz, 2H, NCH$_2$), 5.55 (d, J = 1.1 Hz, 1H, C=CHH), 5.87 (d, J = 1.1 Hz, 1H, C=CHH), 6.91 (s, 1H, 3-H), 7.22 (d, J = 8.6 Hz, 2H, Ar), 7.50 (dd, J = 8.6, 1.1 Hz, 1H, Ar), 7.66 (d, J = 8.6 Hz, 2H, Ar), 7.76 (s, 1H, Ar), 8.23 (d, J = 8.6 Hz, 1H, Ar); ^{13}C NMR (125 MHz, CDCl$_3$) δ 14.1, 20.5, 21.6, 29.4, 52.2, 53.9, 62.6, 109.8, 114.6, 118.0 (q, J = 3.6 Hz), 118.2, 120.6 (q, J = 3.6 Hz), 124.6 (q, J = 272.3 Hz), 125.8 (q, J = 32.4 Hz), 126.4 (2C), 129.4, 130.0 (2C), 131.8, 135.7, 138.9, 142.0, 145.3; MS (FAB) m/z (%): 543 [M–H$^+$ (^{81}Br), 100], 541 [M–H$^+$ (^{79}Br), 90]; HRMS (FAB) calcd for C$_{24}$H$_{25}$BrF$_3$N$_2$O$_2$S [M–H$^+$ (^{79}Br)]: 541.0772; found: 541.0775.

2.2.2.21 2-{[N-(2-Bromoprop-2-en-1-yl)-N-butylamino]methyl}-5-(methoxy-carbonyl)-1-tosylindole (7t)

By a procedure identical to that described for indole **7r**, **1c** (121.3 mg, 0.37 mmol) was converted into **7t** (193.0 mg, 98%) as a colorless solid by the reaction at 80 °C for 3 h: mp 74 °C; ^{1}H NMR (500 MHz, CDCl$_3$) δ 0.88 (t, J = 7.4 Hz, 3H, CH$_2$CH_3), 1.26–1.33 (m, 2H, CH$_2$), 1.40–1.46 (m, 2H, CH$_2$), 2.35 (s, 3H, ArCH$_3$), 2.57–2.60 (m, 2H, NCH$_2$), 3.38 (s, 2H, NCH$_2$), 3.92 (s, 3H, OCH$_3$), 4.05 (d, J = 1.1 Hz, 2H, NCH$_2$), 5.55 (s, 1H, C=CHH), 5.88 (d, J = 1.1 Hz, 1H, C=CHH), 6.90 (s, 1H, 3-H), 7.20 (d, J = 8.6 Hz, 2H, Ar), 7.66 (d, J = 8.6 Hz, 2H, Ar), 7.95 (dd, J = 8.6, 1.7 Hz, 1H, Ar), 8.16–8.19 (m, 2H, Ar); ^{13}C NMR (125 MHz, CDCl$_3$) δ 14.1, 20.5, 21.6, 29.4, 52.1, 52.2, 53.9, 62.6, 110.3, 114.2, 118.2, 122.7, 125.2, 125.5, 126.4 (2C), 129.5, 129.9 (2C), 131.8, 135.8, 140.0, 141.5, 145.2, 167.3; MS (FAB) m/z (%): 533 [M–H$^+$ (^{81}Br), 100], 531 [M–H$^+$ (^{79}Br), 95]; HRMS (FAB) calcd for C$_{25}$H$_{28}$BrN$_2$O$_4$S [M–H$^+$ (^{79}Br)]: 531.0953; found: 531.0957.

2.2.2.22 2-{[N-(2-Bromoprop-2-en-1-yl)-N-butylamino]methyl}-5-methyl-1-tosylindole (7u)

By a procedure identical to that described for indole **7r**, **1d** (105.1 mg, 0.37 mmol) was converted into **7u** (177 mg, 98%) as an yellow oil by the reaction at 80 °C for 3 h, then in the presence of Et$_3$N (205.5 μL, 1.47 mmol) under reflux for 1 h: ^{1}H NMR (500 MHz, CDCl$_3$) δ 0.87 (t, J = 7.2 Hz, 3H, CH$_2$CH_3), 1.25–1.32 (m, 2H, CH$_2$), 1.40–1.46 (m, 2H, CH$_2$), 2.32 (s, 3H, ArCH$_3$), 2.39 (s, 3H, ArCH$_3$), 2.56–2.59 (m, 2H, NCH$_2$), 3.37 (s, 2H, NCH$_2$), 4.03 (d, J = 1.1 Hz, 2H, NCH$_2$),

5.53 (d, $J = 1.1$ Hz, 1H, C=CHH), 5.88 (d, $J = 1.1$ Hz, 1H, C=CHH), 6.73 (s, 1H, 3-H), 7.07 (dd, $J = 8.6$, 1.7 Hz, 1H, Ar), 7.16 (d, $J = 8.6$ Hz, 2H, Ar), 7.24 (s, 1H, Ar), 7.62–7.64 (m, 2H, Ar), 8.00 (d, $J = 8.6$ Hz, 1H, Ar); ^{13}C NMR (125 MHz, CDCl$_3$) δ 14.1, 20.5, 21.2, 21.5, 29.4, 52.2, 53.8, 62.5, 110.2, 114.3, 117.9, 120.5, 125.3, 126.3 (2C), 129.7 (2C), 129.9, 132.0, 133.1, 135.6, 136.0, 139.9, 144.6; MS (FAB) m/z (%): 489 [M–H$^+$ (^{81}Br), 100], 487 [M–H$^+$ (^{79}Br), 90]; HRMS (FAB) calcd for C$_{24}$H$_{28}$BrN$_2$O$_2$S [M–H$^+$ (^{79}Br)]: 487.1055; found: 487.1051.

2.2.2.23 2-{[N-(2-Bromoprop-2-en-1-yl)-N-butylamino]methyl}-1-tosyl-6-(tri-fluoromethyl)indole (7v)

By a procedure identical to that described for of indole **7v**, **1e** (125 mg, 0.37 mmol) was converted into **7s** (188 mg, 94%) as an yellow oil by the reaction at 80 °C, for 3 h then under reflux for 1 h: ^{1}H NMR (500 MHz, CDCl$_3$) δ 0.87 (t, $J = 7.4$ Hz, 3H, CH$_2$CH_3), 1.26–1.33 (m, 2H, CH$_2$), 1.40–1.46 (m, 2H, CH$_2$), 2.36 (s, 3H, ArCH$_3$), 2.57–2.60 (m, 2H, NCH$_2$), 3.38 (s, 2H, NCH$_2$), 4.06 (s, 2H, NCH$_2$), 5.55 (s, 1H, C=CHH), 5.87 (s, 1H, C=CHH), 6.90 (s, 1H, 3-H), 7.22 (d, $J = 8.6$ Hz, 2H, Ar), 7.47 (d, $J = 8.0$ Hz, 1H, Ar), 7.55 (d, $J = 8.0$ Hz, 1H, Ar), 7.65 (d, $J = 8.6$ Hz, 2H, Ar), 8.44 (s, 1H, Ar); ^{13}C NMR (125 MHz, CDCl$_3$) δ 14.0, 20.5, 21.6, 29.4, 52.2, 53.9, 62.7, 109.6, 111.9 (q, $J = 4.8$ Hz), 118.3, 120.3 (m), 120.8, 124.7 (q, $J = 272.3$ Hz), 126.0 (q, $J = 32.4$ Hz), 126.4 (2C), 130.0 (2C), 131.8, 132.2, 135.7, 136.6, 143.0, 145.3; MS (FAB) m/z (%): 543 [M–H$^+$ (^{81}Br), 100], 541 [M–H$^+$ (^{79}Br), 90]; HRMS (FAB) calcd for C$_{24}$H$_{25}$BrF$_3$N$_2$O$_2$S [M–H$^+$ (^{79}Br)]: 541.0772; found: 541.0771.

2.2.2.24 2-{[N-(2-Bromoprop-2-en-1-yl)-N-butylamino]methyl}-6-(methoxy-carbonyl)-1-tosylindole (7w)

By a procedure identical to that described for indole **7r**, **1c** (121 mg, 0.37 mmol) was converted into **7w** (194 mg, 99%) as an yellow oil by the reaction at 80 °C for 3 h, then under reflux for 1.5 h: ^{1}H NMR (500 MHz, CDCl$_3$) δ 0.88 (t, $J = 7.2$ Hz, 3H, CH$_2$CH_3), 1.26–1.33 (m, 2H, CH$_2$), 1.40–1.46 (m, 2H, CH$_2$), 2.34 (s, 3H, ArCH$_3$), 2.57–2.60 (m, 2H, NCH$_2$), 3.38 (s, 2H, NCH$_2$), 3.95 (s, 3H, OCH$_3$), 4.07 (d, $J = 1.1$ Hz, 2H, NCH$_2$), 5.55 (s, 1H, C=CHH), 5.88 (d, $J = 1.1$ Hz, 1H, C=CHH), 6.89 (s, 1H, 3-H), 7.21 (d, $J = 8.6$ Hz, 2H, Ar), 7.49 (d, $J = 8.0$v, 1H, Ar), 7.68 (d, $J = 8.6$ Hz, 2H, Ar), 7.93 (dd, $J = 8.0$, 1.1 Hz, 1H, Ar), 8.85 (s, 1H, Ar); ^{13}C NMR (125 MHz, CDCl$_3$) δ 14.0, 20.4, 21.5, 29.4, 52.1, 52.3, 53.9, 62.7, 109.8, 116.1, 118.2, 120.2, 124.8, 125.7, 126.4 (2C), 129.9 (2C), 131.8, 133.4, 135.8, 136.9, 143.5, 145.1, 167.4; MS (FAB) m/z (%): 533 [M–H$^+$ (^{81}Br), 100], 531 [M–H$^+$ (^{79}Br), 90]; HRMS (FAB) calcd for C$_{25}$H$_{28}$BrN$_2$O$_4$S [M–H$^+$ (^{79}Br)]: 531.0953; found: 531.0947.

2.2.2.25 2-{[N-(2-Bromobenzyl)-N-butylamino]methyl}-1-tosylindole (7x)

By a procedure similar to that described for indole **7a**, **1a** (50.0 mg, 0.18 mmol) was converted into **7x** (77.8 mg, 80%) as a colorless solid by treatment with N-(2-bromobenzyl)butanamine **3h** (49.1 mg, 0.20 mmol) at 80 °C for 3 h, then under reflux for 1 h: mp 72 °C; ^{1}H NMR (400 MHz, CDCl$_3$) δ 0.87 (t, J = 7.3 Hz, 3H, CH$_2$CH$_3$), 1.25–1.34 (m, 2H, CH$_2$), 1.48–1.56 (m, 2H, CH$_2$), 2.31 (s, 3H, ArCH$_3$), 2.57–2.61 (m, 2H, NCH$_2$), 3.76 (s, 2H, NCH$_2$), 4.02 (d, J = 1.0 Hz, 2H, NCH$_2$), 6.77 (s, 1H, 3-H), 7.04–7.26 (m, 6H, Ar), 7.42–7.58 (m, 5H, Ar), 8.13 (d, J = 8.5 Hz, 1H, Ar); ^{13}C NMR (100 MHz, CDCl$_3$) δ 14.1, 20.6, 21.5, 29.4, 52.7, 54.8, 58.5, 110.4, 114.6, 120.4, 123.5, 123.9, 124.0, 126.3 (2C), 127.2, 128.0, 129.7 (2C), 129.8, 129.9, 132.6, 136.0, 137.5, 138.8, 140.3, 144.6; MS (FAB) m/z (%): 525 [M–H$^+$ (^{81}Br),100], 523 [M–H$^+$ (^{79}Br), 95]; HRMS (FAB) calcd for C$_{27}$H$_{28}$BrN$_2$O$_2$S [M–H$^+$ (^{79}Br)]: 523.1055; found: 523.1065.

2.2.3 General Procedure for Synthesis of Tetrahydropyridine-Fused Indole

2.2.3.1 Synthesis of 2-Butyl-4-methylene-9-tosyl-2,3,4,9-tetrahydro-1H-pyrido[3,4-b]indole (9a)

The mixture of indole **7r** (50.0 mg, 0.11 mol), Pd(OAc)$_2$ (2.4 mg, 0.011 mmol), PPh$_3$ (5.5 mg, 0.021 mmol), and CsOAc (40.4 mg, 0.021 mmol) in DMA (2 mL) was stirred at 100 °C for 0.5 h under argon. Concentration under reduced pressure followed by column chromatography purification over silica gel with hexane–AcOEt (4:1) gave **9a** (26.8 mg, 65%) as an yellow oil: ^{1}H NMR (500 MHz, CDCl$_3$) δ 0.94 (t, J = 7.4 Hz, 3H, CH$_2$CH$_3$), 1.30–1.38 (m, 2H, CH$_2$CH$_3$), 1.53–1.59 (m, 2H, NCH$_2$CH$_2$), 2.33 (s, 3H, ArCH$_3$), 2.53–2.56 (m, 2H, NCH$_2$CH$_2$), 3.38 (s, 2H, NCH$_2$), 4.16 (s, 2H, NCH$_2$), 5.10 (s, 1H, C=CHH), 5.59 (s, 1H, C=CHH), 7.20 (d, J = 8.6 Hz, 2H, Ar), 7.26–7.33 (m, 2H, Ar), 7.67 (d, J = 8.6 Hz, 2H, Ar), 7.76–7.78 (m, 1H, Ar), 8.18 (d, J = 8.1 Hz, 1H, Ar); ^{13}C NMR (125 MHz, CDCl$_3$) δ 14.0, 20.6, 21.5, 29.6, 51.5, 55.8, 57.7, 108.6, 114.4, 116.5, 120.4, 124.0, 124.4, 126.4 (2C), 127.3, 130.0 (2C), 135.5, 135.7, 136.1, 136.6, 145.1; MS (FAB) m/z (%): 395 (MH$^+$, 100); HRMS (FAB) calcd for C$_{23}$H$_{27}$N$_2$O$_2$S (MH$^+$): 395.1793; found: 395.1804.

2.2.3.2 2-Butyl-4-methylene-9-tosyl-6-(trifluoromethyl)-2,3,4,9-tetrahydro-1H-pyrido[3,4-b]indole (9b)

By a procedure identical to that described for **9a**, **7s** (57.2 mg, 0.11 mmol) was converted into **9b** (31.3 mg, 64%) as an yellow solid: mp 133 °C; ^{1}H NMR (500 MHz, CDCl$_3$) δ 0.94 (t, J = 7.2 Hz, 3H, CH$_2$CH$_3$), 1.31–1.38 (m, 2H, CH$_2$), 1.52–1.58 (m, 2H, CH$_2$), 2.36 (s, 3H, ArCH$_3$), 2.53–2.56 (m, 2H, NCH$_2$), 3.39 (s,

2H, NCH$_2$), 4.16 (s, 2H, NCH$_2$), 5.15 (s, 1H, C=CHH), 5.59 (s, 1H, C=CHH), 7.24
(d, J = 8.6 Hz, 2H, Ar), 7.56 (dd, J = 8.6, 1.7 Hz, 1H, Ar), 7.68 (d, J = 8.6 Hz,
2H, Ar), 8.01–8.03 (m, 1H, Ar), 8.28 (d, J = 8.6 Hz, 1H, Ar); ^{13}C NMR
(125 MHz, CDCl$_3$) δ 14.0, 20.6, 21.6, 29.6, 51.4, 55.8, 57.5, 109.2, 114.5, 116.3,
117.6 (q, J = 3.6 Hz), 121.2 (q, J = 3.6 Hz), 124.5 (q, J = 271.1 Hz), 126.3 (q,
J = 32.4 Hz), 126.5 (2C), 127.0, 130.2 (2C), 135.3, 135.4, 137.2, 138.1, 145.6;
MS (FAB) m/z (%): 463 (MH$^+$, 100); HRMS (FAB) calcd for C$_{24}$H$_{26}$F$_3$N$_2$O$_2$S
(MH$^+$): 463.1667; found: 463.1671.

2.2.3.3 2-Butyl-4-methylene-6-(methoxycarbonyl)-9-tosyl-2,3,4,9-tetrahydro-1*H*-pyrido[3,4-*b*]indole (9c)

By a procedure identical to that described for **9a**, **7t** (56.1 mg, 0.11 mmol) was
converted into **9c** (25.5 mg, 54%) as an yellow oil: ^{1}H NMR (500 MHz, CDCl$_3$) δ
0.94 (t, J = 7.4 Hz, 3H, CH$_2$CH_3), 1.31–1.38 (m, 2H, CH$_2$), 1.53–1.59 (m, 2H,
CH$_2$), 2.35 (s, 3H, ArCH$_3$), 2.54–2.57 (m, 2H, NCH$_2$), 3.39 (s, 2H, NCH$_2$), 3.93 (s,
3H, OCH$_3$), 4.15 (s, 2H, NCH$_2$), 5.16 (s, 1H, C=CHH), 5.70 (s, 1H, C=CHH), 7.22
(d, J = 8.6 Hz, 2H, Ar), 7.68 (d, J = 8.6 Hz, 2H, Ar), 8.01 (dd, J = 8.6, 1.7 Hz,
1H, Ar), 8.22 (d, J = 8.6 Hz, 1H, Ar), 8.48 (d, J = 1.7 Hz, 1H, Ar); ^{13}C NMR
(125 MHz, CDCl$_3$) δ 14.0, 20.6, 21.6, 29.5, 51.4, 52.2, 55.8, 57.5, 109.4, 114.0,
116.7, 122.4, 125.7, 125.9, 126.4 (2C), 127.1, 130.1 (2C), 135.3 (2C), 136.7,
139.1, 145.5, 167.1; MS (FAB) m/z (%): 453 (MH$^+$, 100); HRMS (FAB) calcd for
C$_{25}$H$_{29}$N$_2$O$_4$S (MH$^+$): 453.1848; found: 453.1854.

2.2.3.4 2-Butyl-6-methyl-4-methylene-9-tosyl-2,3,4,9-tetrahydro-1*H*-pyrido[3,4-*b*]indole (9d)

By a procedure identical to that described for **9a**, **7u** (51.5 mg, 0.11 mmol) was
converted into **9d** (26.5 mg, 62%) as an yellow oil: ^{1}H NMR (500 MHz, CDCl$_3$) δ
0.94 (t, J = 7.2 Hz, 3H, CH$_2$CH_3), 1.30–1.37 (m, 2H, CH$_2$), 1.52–1.58 (m, 2H,
CH$_2$), 2.33 (s, 3H, ArCH$_3$), 2.43 (s, 3H, ArCH$_3$), 2.52–2.55 (m, 2H, NCH$_2$), 3.36
(s, 2H, NCH$_2$), 4.14 (s, 2H, NCH$_2$), 5.08 (s, 1H, C=CHH), 5.58 (s, 1H, C=CHH),
7.12 (dd, J = 8.6, 1.1 Hz, 1H, Ar), 7.19 (J = 8.0 Hz, 2H, Ar), 7.54–7.56 (m, 1H,
Ar), 7.64–7.66 (m, 2H, Ar), 8.04 (d, J = 8.6 Hz, 1H, Ar); ^{13}C NMR (125 MHz,
CDCl$_3$) δ 14.0, 20.6, 21.46, 21.53, 29.6, 51.5, 55.8, 57.8, 108.5, 114.0, 116.4,
120.5, 125.6, 126.4 (2C), 127.5, 129.9 (2C), 133.6, 134.8, 135.6, 135.7, 136.2,
144.9; MS (FAB) m/z (%): 409 (MH$^+$, 100); HRMS (FAB) calcd for C$_{24}$H$_{29}$N$_2$O$_2$S
(MH$^+$): 409.1950; found: 409.1953.

2.2.3.5 2-Butyl-4-methylene-9-tosyl-7-(trifluoromethyl)-2,3,4,9-tetrahydro-1*H*-pyrido[3,4-*b*]indole (9e)

By a procedure identical to that described for **9a**, **7v** (57.2 mg, 0.11 mmol) was
converted into **9e** (30.3 mg, 62%) as an yellow oil: ^{1}H NMR (500 MHz, CDCl$_3$)

δ 0.94 (t, $J = 7.4$ Hz, 3H, CH$_2$CH_3), 1.31–1.38 (m, 2H, CH$_2$), 1.52–1.58 (m, 2H, CH$_2$), 2.36 (s, 3H, ArCH$_3$), 2.53–2.56 (m, 2H, NCH$_2$), 3.38 (s, 2H, NCH$_2$), 4.17 (s, 2H, NCH$_2$), 5.14 (s, 1H, C=CHH), 5.59 (s, 1H, C=CHH), 7.24 (d, $J = 8.6$ Hz, 2H, Ar), 7.53–7.55 (m, 1H, Ar), 7.68 (d, $J = 8.6$ Hz, 2H, Ar), 7.86 (d, $J = 8.0$ Hz, 1H, Ar), 8.48 (s, 1H, Ar); ^{13}C NMR (125 MHz, CDCl$_3$) δ 14.0, 20.5, 21.6, 29.6, 51.4, 55.8, 57.5, 109.2, 111.7 (q, $J = 3.6$ Hz), 116.2, 120.6, 120.8 (q, $J = 3.6$ Hz), 124.5 (q, $J = 272.3$ Hz), 126.4 (q, $J = 32.4$ Hz), 126.5 (2C), 129.7, 130.2 (2C), 135.2, 135.5, 135.8, 138.1, 145.6; MS (FAB) m/z (%): 463 (MH$^+$, 100); HRMS (FAB) calcd for C$_{24}$H$_{26}$F$_3$N$_2$O$_2$S (MH$^+$): 463.1667; found: 463.1665.

2.2.3.6 2-Butyl-7-(methoxycarbonyl)-4-methylene-9-tosyl-2,3,4,9-tetrahydro-1H-pyrido[3,4-b]indole (9f)

By a procedure identical to that described for **9a**, **7w** (56.1 mg, 0.11 mmol) was converted into **9f** (36.8 mg, 77%) as a brown oil: ^{1}H NMR (500 MHz, CDCl$_3$) δ 0.94 (t, $J = 7.2$ Hz, 3H, CH$_2$CH_3), 1.31–1.38 (m, 2H, CH$_2$), 1.52–1.58 (m, 2H, CH$_2$), 2.35 (s, 3H, ArCH$_3$), 2.53–2.56 (m, 2H, NCH$_2$), 3.38 (s, 2H, NCH$_2$), 3.97 (s, 3H, OCH$_3$), 4.17 (s, 2H, NCH$_2$), 5.13 (s, 1H, C=CHH), 5.60 (s, 1H, C=CHH), 7.23 (d, $J = 8.6$ Hz, 2H, Ar), 7.70 (d, $J = 8.6$ Hz, 2H, Ar), 7.80 (d, $J = 8.6$ Hz, 1H, Ar), 7.99 (dd, $J = 8.6, 1.1$ Hz, 1H, Ar), 8.87 (s, 1H, Ar); ^{13}C NMR (125 MHz, CDCl$_3$) δ 14.0, 20.6, 21.6, 29.6, 51.6, 52.3, 55.8, 57.6, 109.1, 115.9, 116.4, 120.0, 125.2, 126.1, 126.5 (2C), 130.1 (2C), 130.8, 135.4, 135.6, 136.0, 138.6, 145.5, 167.1; MS (FAB) m/z (%): 453 (MH$^+$, 100); HRMS (FAB) calcd for C$_{25}$H$_{29}$N$_2$O$_4$S (MH$^+$): 453.1848; found: 453.1839.

2.2.3.7 6-Butyl-8-tosyl-5,6,7,8-tetrahydrobenzo[e]indolo[2,3-c]azepine (10)

The mixture of indole **7x** (50.0 mg, 0.095 mol), Pd(OAc)$_2$ (4.3 mg, 0.019 mmol), PPh$_3$ (10.0 mg, 0.038 mmol), and CsOAc (36.5 mg, 0.19 mmol) in DMA (2 mL) was stirred at 140 °C for 1 h under argon. Concentration under reduced pressure followed by purification by column chromatography with hexane–AcOEt (4:1) gave **10** (42.3 mg, quant) as an yellow oil: ^{1}H NMR (400 MHz, CDCl$_3$) δ 0.98 (t, $J = 7.3$ Hz, 3H, CH$_2$CH_3), 1.39–1.49 (m, 2H, CH$_2$), 1.60–1.68 (m, 2H, CH$_2$), 2.30 (s, 3H, ArCH$_3$), 2.65–2.69 (m, 2H, NCH$_2$), 3.42 (s, 2H, NCH$_2$), 4.05 (s, 2H, NCH$_2$), 7.16 (d, $J = 8.3$ Hz, 2H, Ar), 7.25–7.44 (m, 5H, Ar), 7.66–7.69 (m, 1H, Ar), 7.72–7.74 (m, 1H, Ar), 7.83 (d, $J = 8.3$ Hz, 2H, Ar), 8.31 (d, $J = 8.3$ Hz, 1H, Ar); ^{13}C NMR (125 MHz, CDCl$_3$) δ 14.1, 20.6, 21.5, 30.2, 47.8, 55.8, 56.3, 115.6, 119.4, 123.8, 124.0, 124.8, 126.7 (2C), 127.2, 127.4, 127.6, 128.0, 129.7 (2C), 130.6, 134.4, 135.42, 135.43, 137.02, 137.04, 144.8; MS (FAB) m/z (%): 445 (MH$^+$, 100); HRMS (FAB) calcd for C$_{27}$H$_{29}$N$_2$O$_2$S (MH$^+$): 445.1950; found: 445.1952.

2.2.3.8 *N*-[1-(Naphthalen-1-yl)ethyl]prop-2-en-1-amine (11)

To a stirred solution of 1-(naphthalen-1-yl)ethanamine (1.1 g, 6.42 mmol) and DBU (0.98 mL, 6.55 mmol) in THF was added dropwise allyl bromide (0.56 mL, 6.42 mmol) at rt. The mixture was stirred for 7 h at this temperature, and the whole was extracted with CHCl$_3$. The extract was washed with H$_2$O and dried over MgSO$_4$. Usual workup followed by purification by column chromatography with hexane–AcOEt (1:1) afforded **11** as an yellow oil (779 mg, 57%): ^{1}H NMR (400 MHz, CDCl$_3$) δ 1.49 (d, J = 6.6 Hz, 3H, CHC*H$_3$*), 3.15–3.25 (m, 2H, NCH$_2$), 4.67 (q, J = 6.6 Hz, 1H, NCH), 5.06–5.16 (m, 2H, CH=C*H$_2$*), 5.89–5.98 (m, 1H, C*H*=CH$_2$), 7.44–7.51 (m, 3H, Ar), 7.66 (d, J = 7.1 Hz, 1H, Ar), 7.73 (d, J = 8.0 Hz, 1H, Ar), 7.84–7.87 (m, 1H, Ar), 8.17 (d, J = 8.0 Hz, 1H, Ar); ^{13}C NMR (100 MHz, CDCl$_3$) δ 23.6, 50.3, 52.7, 115.7, 122.6, 122.9, 125.2, 125.7 (2C), 127.1, 128.9, 131.3, 133.9, 137.0, 141.1; MS (FAB) *m/z* (%): 212 (MH$^+$, 100), 196 (55); HRMS (FAB) calcd for C$_{15}$H$_{18}$N (MH$^+$): 212.1439; found: 212.1443.

2.2.3.9 2-{*N*-(Prop-2-en-1-yl)-*N*-[1-(naphthalen-1-yl)ethyl]aminomethyl}-1-tosylindole (12)

By a procedure similar to that described for indole **7a**, **1a** (250 mg, 0.92 mmol) was converted into **12** (539 mg, 85%) as an yellow oil using **11** (214 mg, 1.01 mmol): ^{1}H NMR (400 MHz, CDCl$_3$) δ 1.52 (d, J = 6.6 Hz, 3H, CHC*H$_3$*), 2.19 (s, 3H, ArCH$_3$), 3.33–3.44 (m, 2H, NCH$_2$), 3.99 (d, J = 17.8 Hz, 1H, NC*H*H), 4.20 (d, J = 17.8 Hz, 1H, NCH*H*), 4.84 (q, J = 6.6 Hz, 1H, NCH), 5.09–5.17 (m, 2H, CH=C*H$_2$*), 6.00–6.10 (m, 1H, C*H*=CH$_2$), 6.69 (s, 1H, 3-H), 6.92 (d, J = 8.3 Hz, 2H, Ar), 7.10–7.19 (m, 2H, Ar), 7.32–7.49 (m, 6H, Ar), 7.63 (d, J = 7.3 Hz, 1H, Ar), 7.68 (d, J = 8.3 Hz, 1H, Ar), 7.79 (d, J = 8.3 Hz, 1H, Ar), 8.07 (d, J = 8.0 Hz, 1H, Ar), 8.37 (d, J = 8.3 Hz, 1H, Ar); ^{13}C NMR (100 MHz, CDCl$_3$) δ 16.5, 21.4, 48.9, 54.9, 56.4, 110.2, 114.4, 117.7, 120.2, 123.3, 123.6, 124.18, 124.21, 125.1, 125.3, 125.5, 126.1 (2C), 127.4, 128.6, 129.5 (2C), 129.8, 131.8, 133.9, 135.4, 135.6, 137.2, 140.0, 141.2, 144.3; MS (FAB) *m/z* (%): 495 (MH$^+$, 100), 479 (50), 339 (30), 284 (60); HRMS (FAB) calcd for C$_{31}$H$_{31}$N$_2$O$_2$S (MH$^+$): 495.2106; found: 495.2108.

2.2.3.10 Calindol (13)

To a stirred mixture of Pd(PPh$_3$)$_4$ (8.9 mg, 0.0077 mmol) and 1,3-dimthylbarbituric acid (179.6 mg, 1.15 mmol) in CH$_2$Cl$_2$ (4 mL) was added a solution of **12** (190.0 mg, 0.38 mmol) in CH$_2$Cl$_2$ (1 mL) at rt under argon. The reaction mixture was stirred at 40 °C for 1 h, and the whole was extracted with CHCl$_3$. The extract was washed successively with Na$_2$CO$_3$ and H$_2$O. Usual workup followed by purification by column chromatography with hexane–EtOAc (3:1) afforded *N*-tosylcalindole **13a** (156.9 mg, 90%) as a colorless solid: mp 104 °C; ^{1}H NMR

(400 MHz, CDCl$_3$) δ 1.48 (d, J = 6.6 Hz, 3H, CHCH_3), 2.26 (s, 3H, ArCH$_3$), 2.36 (br s, 1H, NH), 3.95 (d, J = 15.4 Hz, 1H, NCHH), 4.15 (d, J = 15.4 Hz, 1H, NCHH), 4.66 (q, J = 6.6 Hz, 1H, NCH), 6.36 (s, 1H, 3-H), 7.03 (d, J = 8.3 Hz, 2H, Ar), 7.20–7.31 (m, 2H, Ar), 7.38–7.51 (m, 4H, Ar), 7.56 (d, J = 8.3 Hz, 2H, Ar), 7.75–7.78 (m, 2H, Ar), 7.87–7.89 (m, 1H, Ar), 8.07 (d, J = 8.0 Hz, 1H, Ar), 8.17 (d, J = 8.3 Hz, 1H, Ar); ^{13}C NMR (100 MHz, CDCl$_3$) δ 21.5, 23.8, 44.9, 51.7, 111.2, 114.7, 120.6, 123.0, 123.1, 123.6, 124.4, 125.3, 125.7, 125.8, 126.2 (2C), 127.2, 128.9, 129.4, 129.7 (2C), 131.3, 134.0, 135.7, 137.4, 139.6, 140.6, 144.7; MS (FAB) m/z (%): 455 (MH$^+$, 100), 479 (20), 284 (50); HRMS (FAB) calcd for C$_{28}$H$_{27}$N$_2$O$_2$S (MH$^+$): 455.1793; found: 455.1787.

The mixture of N-tosylated indole **13a** (110 mg, 0.24 mmol) and TBAF (1 M in THF, 4.8 mL, 4.8 mmol) was stirred under reflux for 3 h. The whole was extracted with Et$_2$O, and the extract was washed with H$_2$O. Usual workup followed by purification by column chromatography with hexane–EtOAc (1:1) yielded calindol **13** as a brown oil (72.8 mg, quant): ^{1}H NMR (400 MHz, CDCl$_3$) δ 1.53 (d, J = 6.6 Hz, 3H, CHCH_3), 2.21 (br s, 1H, NH), 3.84 (d, J = 14.1 Hz, 1H, NCHH), 3.90 (d, J = 14.1 Hz, 1H, NCHH), 4.70 (q, J = 6.6 Hz, 1H, NCH), 6.27 (s, 1H, 3-H), 7.05–7.09 (m, 1H, Ar), 7.12–7.16 (m, 1H, Ar), 7.30 (d, J = 7.8 Hz, 1H, Ar), 7.46–7.54 (m, 4H, Ar), 7.68 (d, J = 7.1 Hz, 1H, Ar), 7.77 (d, J = 8.3 Hz, 1H, Ar), 7.86–7.96 (m, 1H, Ar), 8.10–8.13 (m, 1H, Ar), 8.44 (br s, 1H, 1-H); ^{13}C NMR (100 MHz, CDCl$_3$) δ 23.4, 44.8, 53.0, 100.1, 110.7, 119.6, 120.1, 121.4, 122.6, 122.9, 125.5, 125.7, 125.9, 127.5, 128.5, 129.0, 131.3, 134.0, 135.8, 137.8, 140.5; MS (FAB) m/z (%): 401 (MH$^+$, 100); HRMS (FAB) calcd for C$_{21}$H$_{21}$N$_2$ (MH$^+$): 301.1715; found: 301.1716.

2.2.3.11 2-Ethynyl-N-methylbenzenesulfonamide (14a)

To a solution of 2-bromobenzenesulfonylchloride **S8** (3.00 g, 11.8 mmol) in CHCl$_3$ (100 mL) was added dropwise methanamine (40% in MeOH, 3.47 mL, 33.5 mmol) at 0 °C and the reaction mixture was stirred at rt for 5 min. After concentration under reduced pressure, the residue was dissolved in Et$_2$O. The solution was washed successively with 1 N HCl and brine, and dried over MgSO$_4$. The filtrate was concentrated under reduced pressure and the residue was purified by column chromatography over silica gel with hexane–EtOAc (3:1) to give the known sulfonamide **S9a** (2.70 g, 92%).

To a stirred mixture of **S9a** (2.65 g, 10.7 mmol), PdCl$_2$(PPh$_3$)$_2$ (0.38 g, 0.53 mmol) and CuI (0.10 g, 0.53 mmol) in a mixed solvent of THF (25 mL) and Et$_3$N (25 mL) was added TMS-acetylene (1.75 mL, 12.8 mmol) at rt under argon, and the reaction mixture was stirred at 100 °C for 2 h. The mixture was filtered through a pad of Celite. The filtrate was concentrated under reduced pressure and the residue was purified by column chromatography over silica gel with hexane–EtOAc (5:1) to give **S10a** as an yellow oil (2.63 g, 92%). To a solution of **S10a** (54.0 mg, 0.20 mmol) in THF (1 mL) was added TBAF (1 M in THF, 0.21 mL, 0.21 mmol) at −78 °C and the reaction mixture was stirred for 1 min at this temperature. After quenching with aqueous saturated citric acid, the whole was extracted with Et$_2$O. The extract was washed with water, NaHCO$_3$, and brine, and dried over MgSO$_4$. Usual workup followed by purification by column chromatography over silica gel with hexane–EtOAc (3:1) gave **14a** (33.6 mg, 86%) as a pale yellow solid, which was recrystallized from hexane–CHCl$_3$ to give pure **14a** as pale yellow crystals: mp 94 °C; IR (neat) cm^{-1} 3268 cm^{-1} (NH), 2110 (C≡C); ^{1}H NMR (500 MHz, CDCl$_3$) δ 2.63 (d, J = 5.2 Hz, 3H, CH$_3$), 3.65 (s, 1H, C≡CH), 5.15–5.18 (m, 1H, NH), 7.51–7.57 (m, 2H, Ar), 7.70 (d, J = 7.4 Hz, 1H, Ar), 8.05–8.07 (m, 1H, Ar); ^{13}C NMR (125 MHz, CDCl$_3$) δ 29.4, 80.1, 85.7, 119.3, 129.2, 129.7, 132.3, 135.2, 140.3. Anal. Calcd for C$_9$H$_9$NO$_2$S: C, 55.37; H, 4.65; N, 7.17. Found: C, 55.41; H, 4.65; N, 7.16.

2.2.3.12 *N*-Ethyl-2-ethynylbenzenesulfonamide (14b)

By a procedure similar to that described for **S9a**, **S8** (2.00 g, 7.83 mmol) was converted into the known sulfonamide **S9b** (1.90 g, 92%) using ethylamine (70% in H$_2$O, 1.82 mL, 22.3 mmol).

By a procedure identical to that described for **S10a**, **S9b** (820 mg, 3.12 mmol) was converted into **S10b** as a brown oil (568 mg, 65%). By a procedure identical to that described for **14a**, **S10b** (360 mg, 1.28 mmol) was converted into **14b** (210 mg, 79%): brown crystals; mp 98 °C; IR (neat) cm^{-1} 3293 (NH), 2109 (C≡C); ^{1}H NMR (500 MHz, CDCl$_3$) δ 1.10 (t, J = 7.4 Hz, 3H, CH$_3$), 2.95–3.01 (m, 2H, CH$_2$), 3.69 (s, 1H, C≡CH), 5.23 (t, J = 5.4 Hz, 1H, NH), 7.49–7.56 (m, 2H, Ar), 7.69 (dd, J = 7.4, 1.1 Hz, 1H, Ar), 8.05 (m, J = 7.4, 1.1 Hz, 1H, Ar); ^{13}C NMR (125 MHz, CDCl$_3$) δ 15.0, 38.4, 80.3, 85.9, 119.3, 129.18, 129.22, 132.1, 135.2, 141.6. Anal. Calcd for C$_{10}$H$_{11}$NO$_2$S: C, 57.39; H, 5.30; N, 6.69. Found: C, 57.31; H, 5.37; N, 6.64.

2.2.3.13 2-Ethynyl-*N*-*p*-tolylbenzenesulfonamide (14c)

To a solution of 2-bromobenzenesulfonyl chloride **S8** (1.00 g, 3.92 mmol) in DMF (50 mL) was added *p*-toluidine (1.68 g, 15.7 mmol) at 0 °C and the reaction mixture was stirred at rt for 10 min. The whole was extracted with Et$_2$O. The extract was washed successively with 1 N HCl and brine, and dried over MgSO$_4$. The filtrate was concentrated under reduced pressure and the residue was purified by column chromatography over silica gel with hexane–EtOAc (5:1) to give **S9c** (1.03 g, 81%) as a colorless solid, which was recrystallized from hexane–CHCl$_3$ to give pure **S9c** as colorless crystals: mp 151–152 °C; IR (neat) cm^{-1} 3284 (NH); ^{1}H NMR (500 MHz, CDCl$_3$) δ 2.23 (s, 3H, ArCH$_3$), 6.99–7.02 (m, 4H, Ar), 7.07 (br s, 1H, NH), 7.33–7.37 (m, 2H, Ar), 7.69–7.72 (m, 1H, Ar), 7.97–8.01 (m, 1H, Ar); ^{13}C NMR (125 MHz, CDCl$_3$) δ 20.8, 119.6, 122.2 (2C), 127.7, 129.8 (2C), 132.2, 132.9, 133.9, 134.9, 135.7, 137.8. Anal. Calcd for C$_{13}$H$_{12}$BrNO$_2$S: C, 47.86; H, 3.71; N, 4.29. Found: C, 47.79; H, 3.78; N, 4.25.

By a procedure identical to that described for **S10a**, **S9c** (944 mg, 2.90 mmol) was converted into **S10c** (722 mg, 73%) as an yellow oil. By a procedure identical to that described for **14a**, **S10c** (671 mg, 1.96 mmol) was converted into **14c** as a white solid (283 mg, 53%), which was recrystallized from hexane–CHCl$_3$ to give pure **14c** as colorless crystals: mp 158 °C; IR (neat) cm^{-1} 3290 (NH), 2110 (C≡C); ^{1}H NMR (500 MHz, CDCl$_3$) δ 2.22 (s, 3H, ArCH$_3$), 3.77 (s, 1H, C≡CH), 6.98–7.03 (m, 4H, Ar), 7.19 (br s, 1H, NH), 7.36–7.39 (m, 1H, Ar), 7.45–7.48 (m, 1H, Ar), 7.66 (d, J = 8.0 Hz, 1H, Ar), 7.89 (d, J = 8.0 Hz, 1H, Ar); ^{13}C NMR (125 MHz, CDCl$_3$) δ 20.8, 80.7, 86.0, 119.4, 122.5 (2C), 129.1, 129.8 (3C), 132.4, 133.2, 135.1, 135.7, 140.6. Anal. Calcd for C$_{15}$H$_{13}$NO$_2$S: C, 66.40; H, 4.83; N, 5.16. Found: C, 66.27; H, 4.86; N, 5.27.

2.2.3.14 2-Ethynyl-*N*-phenylbenzenesulfonamide (14d)

By a procedure identical to that described for **S9c**, **S8** (3.00 g, 11.8 mmol) was converted into the known compound **S9d** (2.87 g, 79%) using aniline (3.05 mL, 33.5 mmol).

By a procedure identical to that described for **S10a**, **S9d** (2.50 g, 8.04 mmol) was converted into **S10d** as an yellow oil (2.43 g, 97%). By a procedure identical to that described for **14a**, **S10d** (55.0 mg, 0.18 mmol) was converted into **14d** (33.0 mg, 71%): brown crystals; mp 107 °C; IR (neat) cm^{-1} 3283 cm^{-1} (NH), 2111 (C≡C); ^{1}H NMR (500 MHz, CDCl$_3$) δ 3.78 (s, 1H, C≡CH), 7.05–7.08 (m, 1H, Ar), 7.13–7.15 (m, 2H, Ar), 7.18–7.21 (m, 2H, Ar), 7.33 (br s, 1H, NH), 7.37–7.40 (m, 1H, Ar), 7.45–7.48 (m, 1H, Ar), 7.65 (dd, J = 7.4, 1.1 Hz, 1H, Ar), 7.93 (dd, J = 8.0, 1.1 Hz, 1H, Ar); ^{13}C NMR (125 MHz, CDCl$_3$) δ 80.5, 86.1, 119.5, 121.8 (2C), 125.6, 129.1, 129.2 (2C), 129.8, 132.5, 135.1, 136.0, 140.5. Anal. Calcd for C$_{14}$H$_{11}$NO$_2$S: C, 65.35; H, 4.31; N, 5.44. Found: C, 65.43; H, 4.46; N, 5.53.

2.2.3.15 2-Ethynyl-*N*-(4-methoxyphenyl)benzenesulfonamide (14e)

By a procedure identical to that described for **S9c**, **S8** (1.00 g, 3.92 mmol) was converted into **S9e** (1.13 g, 84%) using *p*-anisidine (1.45 g, 11.7 mmol): colorless crystals; mp 127–128 °C; IR (neat) cm^{-1} 3285 (NH); ^{1}H NMR (500 MHz, CDCl$_3$) δ 3.71 (s, 3H, OCH$_3$), 6.72 (d, J = 8.6 Hz, 2H, Ar), 7.05 (d, J = 8.6 Hz, 2H, Ar), 7.09 (br s, 1H, Ar), 7.31–7.37 (m, 2H, Ar), 7.72 (dd, J = 7.4, 1.1 Hz, 1H, Ar), 7.92 (dd, J = 7.4, 2.3 Hz, 1H, Ar); ^{13}C NMR (125 MHz, CDCl$_3$) δ 55.3, 114.4 (2C), 119.6, 125.3 (2C), 127.8, 128.0, 132.2, 133.9, 134.9, 137.7, 158.1. Anal. Calcd for C$_{13}$H$_{12}$BrNO$_3$S: C, 45.63; H, 3.53; N, 4.09. Found: C, 45.78; H, 3.49; N, 4.15.

By a procedure identical to that described for **S10a**, **S9e** (1.03 g, 3.02 mmol) was converted into **S10e** as an yellow oil (778 mg, 72%). By a procedure identical to that described for **14a**, **S10e** (685 mg, 1.91 mmol) was converted into **14e** (364 mg, 66%): pale yellow crystals; mp 122 °C; IR (neat) cm^{-1} 3291 (NH), 2254 cm^{-1} (C≡C); ^{1}H NMR (500 MHz, CDCl$_3$) δ 3.71 (s, 3H, OCH$_3$), 3.78 (s, 1H, C≡CH), 6.71 (d, J = 8.6 Hz, 2H, Ar), 7.06 (d, J = 8.6 Hz, 2H, Ar), 7.14 (br s, 1H, NH), 7.35–7.38 (m, 1H, Ar), 7.46–7.49 (m, 1H, Ar), 7.68 (d, J = 8.0 Hz, 1H, Ar), 7.84 (d, J = 8.0 Hz, 1H, Ar); ^{13}C NMR (125 MHz, CDCl$_3$) δ 55.3, 80.6, 86.2, 114.3 (2C), 119.4, 125.3 (2C), 128.3, 129.1, 129.8, 132.3, 135.0, 140.6, 158.0. Anal. Calcd for C$_{15}$H$_{13}$NO$_3$S: C, 62.70; H, 4.56; N, 4.87. Found: C, 62.77; H, 4.53; N, 4.96.

2.2.3.16 *N*-(4-Chlorophenyl)-2-ethynylbenzenesulfonamide (14f)

By a procedure identical to that described for **S9c**, **S8** (1.00 g, 3.92 mmol) was converted into **S9f** (1.06 g, 78%) using 4-chloroaniline (1.99 g, 15.7 mmol): colorless crystals; mp 133–134 °C; IR (neat) cm^{-1} 3276 (NH); ^{1}H NMR (500 MHz, CDCl$_3$) δ 7.08–7.10 (m, 2H, Ar), 7.14–7.17 (m, 2H, Ar), 7.35–7.40 (m, 2H, Ar), 7.49 (br s, 1H, NH), 7.67–7.70 (m, 1H, Ar), 8.01–8.05 (m, 1H, Ar); ^{13}C NMR (125 MHz, CDCl$_3$) δ 119.6, 122.8 (2C), 127.8, 129.4 (2C), 131.1, 132.3, 134.2, 134.3, 135.1, 137.3. Anal. Calcd for C$_{12}$H$_9$BrClNO$_2$S: C, 41.58; H, 2.62; N, 4.04. Found: C, 41.54; H, 2.62; N, 4.04.

By a procedure identical to that described for **S10a**, **S9f** (0.93 g, 2.71 mmol) was converted into **S10f** (546 mg, 55%) as an yellow oil. By a procedure identical to that described for **14a**, **S10f** (491 mg, 1.35 mmol) was converted into **14f** (238 mg, 61%): yellow crystals; mp 123–124 °C; IR (neat) cm^{-1} 3286 (NH), 2112 (C≡C); ^{1}H NMR (500 MHz, CDCl$_3$) δ 3.79 (s, 1H, C≡CH), 7.07–7.10 (m, 2H, Ar), 7.15–7.18 (m, 2H, Ar), 7.37 (br s, 1H, NH), 7.40–7.43 (m, 1H, Ar), 7.48–7.51 (m, 1H, Ar), 7.66 (d, J = 8.0 Hz, 1H, Ar), 7.92 (d, J = 8.0 Hz, 1H, Ar); ^{13}C NMR (125 MHz, CDCl$_3$) δ 80.4, 86.3, 119.4, 123.2 (2C), 129.2, 129.4 (2C), 129.8, 131.2, 132.7, 134.5, 135.2, 140.1. Anal. Calcd for C$_{14}$H$_{10}$ClNO$_2$S: C, 57.63; H, 3.45; N, 4.80. Found: C, 57.79; H, 3.64; N, 4.80.

2.2.4 General Procedure for Synthesis of Benzo[e][1,2]thiazine-1,1-dioxide

2.2.4.1 3-[(*N,N*-Diisopropylamino)methyl]-2-methyl-2*H*-benzo[*e*][1,2]thiazine-1,1-dioxide (15a)

To a stirred mixture of **14a** (50.0 mg, 0.26 mmol), (HCHO)$_n$ (15.4 mg, 0.51 mmol), and CuBr (1.8 mg, 0.013 mmol) in dioxane (3 mL) was added diisopropylamine (43.1 µL, 0.31 mmol) at rt under argon. The reaction mixture was stirred at 100 °C for 16 h. Concentration under reduced pressure followed by column chromatography purification over silica gel with hexane–EtOAc (8:1) gave **15a** as a pale yellow solid (27.2 mg, 34%): mp 94.5–98.5 °C; ^{1}H NMR (500 MHz, CDCl$_3$) δ 1.06 (d, J = 6.9 Hz, 12H, 4 × CHCH_3), 3.11–3.19 (m, 2H, 2 × CH), 3.40 (s, 3H, NCH$_3$), 3.50 (s, 2H, NCH$_2$), 6.51 (s, 1H, 4-H), 7.32 (d, J = 8.0 Hz, 1H, Ar), 7.39–7.42 (m, 1H, Ar), 7.52–7.55 (m, 1H, Ar), 7.84 (d, J = 8.0 Hz, 1H Ar); ^{13}C NMR (125 MHz, CDCl$_3$) δ 20.4 (4C), 30.9, 47.5 (2C), 48.4, 109.0, 121.5, 126.3, 126.8, 130.5, 131.7, 132.9, 143.5. Anal. Calcd for C$_{16}$H$_{24}$N$_2$O$_2$S: C, 62.30; H, 7.84; N, 9.08. Found: C, 62.09; H, 7.57; N, 8.88.

2.2.4.2 3-[(*N,N*-Diisopropylamino)methyl]-2-ethyl-2*H*-benzo[*e*][1,2]thiazine-1,1-dioxide (15b)

By a procedure identical to that described for **15a**, **14b** (25.0 mg, 0.12 mmol) was converted into **15b** as an yellow oil (14.3 mg, 37%): ^{1}H NMR (500 MHz, CDCl$_3$) δ 1.06 (d, J = 6.3 Hz, 12H, 4 × CHCH_3), 1.09 (t, J =7.2 Hz, 3H, CH$_2$CH_3), 3.11–3.18 (m, 2H, 2 × NCH), 3.49 (s, 2H, NCH$_2$), 3.95 (q, J = 7.2 Hz, 2H, CH_2CH$_3$), 6.66 (s, 1H, 4-H), 7.33 (d, J = 8.0 Hz, 1H, Ar), 7.40–7.43 (m, 1H, Ar), 7.51–7.54 (m, 1H, Ar), 7.83 (d, J = 8.0 Hz, 1H, Ar); ^{13}C NMR (125 MHz, CDCl$_3$) δ 15.3, 20.4 (4C), 40.4, 47.7 (2C), 47.9, 110.9, 121.3, 126.5, 127.0, 131.6 (2C), 132.8, 142.9; MS (FAB) *m/z* (%): 323 (MH$^+$, 100); HRMS (FAB) calcd for C$_{17}$H$_{27}$N$_2$O$_2$S (MH$^+$): 323.1793; found, 323.1765.

2.2.4.3 3-[(*N,N*-Diisopropylamino)methyl]-2-(*p*-tolyl)-2*H*-benzo[*e*][1,2]thiazine-1,1-dioxide (15c)

By a procedure identical to that described for **15a** from **14a**, **14c** (25.0 mg, 0.09 mmol) was converted into **15c** (31.9 mg, 90%): colorless crystals; mp 100–101 °C; ^{1}H NMR (500 MHz, CDCl$_3$) δ 0.90 (d, J = 6.3 Hz, 12H, 4 × CH$_3$), 2.34 (s, 3H, ArCH$_3$), 3.01–3.08 (m, 2H, 2 × NCH), 3.21 (d, J = 1.1 Hz, 2H, NCH$_2$), 6.92 (s, 1H, 4-H), 7.06 (d, J = 8.6 Hz, 2H, Ar), 7.14 (d, J = 8.6 Hz, 2H, Ar), 7.40–7.44 (m, 2H, Ar), 7.56–7.59 (m, 1H, Ar), 7.78 (d, J = 8.0 Hz, 1H, Ar); ^{13}C NMR (125 MHz, CDCl$_3$) δ 20.5 (4C), 21.3, 47.8, 48.2 (2C), 111.6, 122.4, 127.0, 127.4, 128.1 (2C), 129.6 (2C), 131.4, 132.0, 133.1, 133.2, 138.5, 145.0.

Anal. Calcd for $C_{22}H_{28}N_2O_2S$: C, 68.72; H, 7.34; N, 7.29. Found: C, 68.45; H, 7.46; N, 7.13.

2.2.4.4 3-[(*N*,*N*-Diisopropylamino)methyl]-2-phenyl-2*H*-benzo[*e*][1,2]thia-zine-1,1-dioxide (15d)

By a procedure identical to that described for **15a** from **14a**, **14d** (25.0 mg, 0.10 mmol) was converted into **15d** (33.0 mg, 92%): colorless crystals; mp 73.5–74.0 °C; ^{1}H NMR (500 MHz, CDCl$_3$) δ 0.88 (d, J = 6.3 Hz, 12H, 4 × CH$_3$), 3.00–3.08 (m, 2H, 2 × NCH), 3.23 (s, 2H, NCH$_2$), 6.92 (s, 1H, 4-H), 7.18–7.20 (m, 2H, Ar), 7.30–7.36 (m, 3H, Ar), 7.41–7.45 (m, 2H, Ar), 7.57–7.60 (m, 1H, Ar), 7.79 (d, J = 7.4 Hz, 1H, Ar); ^{13}C NMR (125 MHz, CDCl$_3$) δ 20.4 (4C), 47.8, 47.9 (2C), 112.0, 122.4, 127.1, 127.5, 128.28 (2C), 128.31, 128.9 (2C), 131.6, 132.1, 133.0, 135.9, 144.8. Anal. Calcd for $C_{21}H_{26}N_2O_2S$: C, 68.08; H, 7.07; N, 7.56. Found: C, 67.96; H, 7.22; N, 7.26.

2.2.4.5 3-[(*N*,*N*-Diisopropylamino)methyl]-2-(4-methoxyphenyl)-2*H*-benzo[*e*][1,2]thiazine-1,1-dioxide (15e)

By a procedure identical to that described for **15a** from **14a**, **14e** (25.0 mg, 0.09 mmol) was converted into **15e** as an yellow oil (31.1 mg, 89%): ^{1}H NMR (500 MHz, CDCl$_3$) δ 0.91 (d, J = 6.9 Hz, 12H, 4 × CHC*H*$_3$), 3.00–3.08 (m, 2H, 2 × NCH), 3.20 (s, 2H, NCH$_2$), 3.79 (s, 3H, OCH$_3$), 6.85 (d, J = 9.2 Hz, 2H, Ar), 6.88 (s, 1H, 4-H), 7.10 (d, J = 9.2 Hz, 2H, Ar), 7.41–7.44 (m, 2H, Ar), 7.56–7.59 (m, 1H, Ar), 7.79 (d, J = 7.4 Hz, 1H, Ar); ^{13}C NMR (125 MHz, CDCl$_3$) δ 20.5 (4C), 47.8, 48.1 (2C), 55.5, 112.2, 114.2 (2C), 122.4, 127.0, 127.3, 128.5, 129.6 (2C), 131.3, 132.0, 133.1, 145.1, 159.6; MS (FAB) *m/z* (%): 401 (MH$^+$, 65); HRMS (FAB) calcd for $C_{22}H_{29}N_2O_3S$ (MH$^+$): 401.1899; found, 401.1893.

2.2.4.6 2-(4-Chlorophenyl)-3-[(*N*,*N*-diisopropylamino)methyl]-2*H*-benzo[*e*][1,2]thiazine-1,1-dioxide (15f)

By a procedure identical to that described for **15a** from **14a**, **14f** (25.0 mg, 0.09 mmol) was converted into **15f** (33.1 mg, 95%): colorless crystals; mp 113–114 °C; ^{1}H NMR (500 MHz, CDCl$_3$) δ 0.89 (d, J = 6.9 Hz, 12H, 4 × CH$_3$), 3.01–3.09 (m, 2H, 2 × NCH), 3.22 (d, J = 1.1 Hz, 2H, NCH$_2$), 6.91 (s, 1H, 4-H), 7.10–7.13 (m, 2H, Ar), 7.30–7.33 (m, 2H, Ar), 7.43–7.46 (m, 2H, Ar), 7.58–7.61 (m, 1H, Ar), 7.78 (d, J = 7.4 Hz, 1H, Ar); ^{13}C NMR (125 MHz, CDCl$_3$) δ 20.4 (4C), 47.8, 47.9 (2C), 112.7, 122.5, 127.2, 127.7, 129.1 (2C), 129.4 (2C), 131.5, 132.3, 132.8, 134.2, 134.5, 144.3; MS (FAB) *m/z* (%): 405 (MH$^+$, 72). Anal. Calcd for $C_{21}H_{25}ClN_2O_2S$: C, 62.28; H, 6.22; N, 6.92. Found: C, 62.15; H, 6.31; N, 6.87.

2.2.4.7 Dimethyl 2-(2-Iodophenyl)malonate (S12)

To a solution of NaH (0.80 g, 20.1 mmol) in $C(O)(OMe)_2$ (15 mL) was added **S11** (1.38 g, 5.01 mmol) at 0 °C. The reaction mixture was stirred at rt for 1.5 h and additional 0.5 h under reflux. To a mixture was added saturated aqueous NH_4Cl at 0 °C, and the mixture was stirred for 10 min. Then water was added and the mixture was extracted with CH_2Cl_2 three times. The organic layer was dried over $MgSO_4$. Usual workup followed by purification by column chromatography over silica gel with hexane–EtOAc (5:1) gave the known compound **S12** (1.38 g, 83%).

2.2.4.8 Dimethyl 2-(2-Ethynylphenyl)malonate (16)

To a stirred solution of **S12** (1.26 g, 3.77 mmol), $PdCl_2(PPh_3)_2$ (66.6 mg, 0.094 mmol) and CuI (17.9 mg, 0.094 mmol) in a mixed solvent of THF (2 mL) and Et_3N (25 mL) was added TMS-acetylene (0.62 mL, 4.52 mmol) at rt under argon, and the reaction mixture was stirred at 55 °C for 20 min. The mixture was filtered through a pad of Celite. The filtrate was concentrated under reduced pressure and the residue was purified by column chromatography over silica gel with hexane–EtOAc (5:1) to give **S13** as a brown oil (1.13 g, 99%). To a solution of **S13** (0.89 g, 2.90 mmol) in THF (10 mL) was added TBAF (1 mol/L in THF, 3.05 mL, 3.05 mmol) at −78 °C and the reaction mixture was stirred for 25 min at this temperature. After the reaction mixture was quenched with aqueous saturated citric acid, the whole was extracted with Et_2O. The extract was washed successively with water, aqueous saturated $NaHCO_3$, and brine, and dried over $MgSO_4$. Usual workup followed by purification by column chromatography over silica gel with hexane–EtOAc (8:1) gave **16** (509.2 mg, 76%) as a red solid which was recrystallized from hexane–$CHCl_3$ to give pure **16** as pink crystals: mp 41–42 °C; IR (neat) cm^{-1} 2106 (C≡C), 1733 cm^{-1} (C=O); ^1H NMR (500 MHz, $CDCl_3$) δ 3.32 (s, 1H, C≡CH), 3.76, (s, 6H, 2 × OMe), 5.37 (s, 1H, ArCH), 7.28–7.31 (m, 1H, Ar), 7.36–7.40 (m, 1H, Ar), 7.50–7.54 (m, 2H, Ar); ^{13}C NMR (125 MHz, $CDCl_3$) δ 52.3 (2C), 55.0, 81.0, 82.3, 122.6, 128.0, 128.8, 129.2, 132.8, 134.9, 168.3 (2C). Anal. Calcd for $C_{13}H_{12}O_4$: C, 67.23; H, 5.21. Found: C, 67.14; H, 5.24.

2.2.4.9 Dimethyl 2-[(*N*,*N*-Diisopropylamino)methyl]indene-1,1-dicarboxylate (17)

To a stirred mixture of **16** (51.0 mg, 0.22 mmol), (HCHO)$_n$ **2a** (13.2 mg, 0.44 mmol), and CuBr (1.58 mg, 0.011 mmol) in DMF (2 mL) was added **3a** diisopropylamine (34.0 μL, 0.24 mmol) at rt under argon. After the reaction mixture was stirred at 110 °C for 30 min, diisopropylethylamine (77.0 μL, 0.44 mmol) was added to the mixture. The mixture was additionally stirred at 110 °C for 9.5 h. Concentration under reduced pressure followed by purification by column chromatography over alumina with hexane–EtOAc (20:1) gave **17** (52.8 mg, 70%) as a brown oil: IR (neat) 1732 (C=O) cm^{-1}; ^{1}H NMR (500 MHz, CDCl$_3$) δ 1.02 (d, J = 14.0 Hz, 12H, 4 × CCH$_3$), 3.06–3.11 (m, 2H, 2 × NCH), 3.50 (d, J = 1.1 Hz, 2H, NCH$_2$), 3.73 (s, 6H, 2 × OCH$_3$), 7.00 (s, 1H, 3-H), 7.15–7.18 (m, 1H, Ar), 7.24–7.31 (m, 2H, Ar), 7.57 (d, J = 7.4 Hz, 1H, Ar); ^{13}C NMR (125 MHz, CDCl$_3$) δ 20.7 (4C), 43.8, 48.7, 53.0 (2C), 70.3, 120.8, 124.8, 125.1, 128.7, 131.6, 141.0, 144.3, 149.4, 168.9 (2C); MS (FAB) *m/z*: 346 (MH$^+$, 100), 286 (35), 245 (60); HRMS (FAB) calcd for C$_{20}$H$_{28}$NO$_4$ (MH$^+$), 346.2018; found, 346.2008.

References

1. Gommermann N, Koradin C, Polborn K, Knochel P (2003) Angew Chem Int Ed 42:5763–5766
2. Gommerman N, Knochel P (2004) Chem Commun 2324–2325
3. Gommerman N, Knochel P (2005) Chem Commun 4175–4177
4. Knöpfel TF, Aschwanden P, Ichikawa T, Watanabe T, Carreira EM (2004) Angew Chem Int Ed 43:5971–5973
5. Aschwanden P, Stephenson CRJ, Carreira EM (2006) Org Lett 8:2437–2440
6. Espada A, Jiménez C, Debitus C, Riguera R (1993) Tetrahedron Lett 34:7773–7776
7. Rashid MA, Gustafson KR, Boyd MRJ (2001) Nat Chem 64:1454–1456
8. Glennon RA, Grella B, Tyacke RJ, Lau A, Westaway J, Hudson AL (2004) Bioorg Med Chem Lett 14:999–1002
9. Liu C, Masuno MN, MacMillan JB, Molinski TF (2004) Angew Chem Int Ed 43:5941–5945
10. Sonnenschein RN, Farias JJ, Tenney K, Mooberry SL, Lobkovsky E, Clardy J, Crews P (2004) Org Lett 6:779–782
11. Kusama H, Takaya J, Iwasawa N (2002) J Am Chem Soc 124:11592–11593
12. Bandini M, Melloni A, Piccinelli F, Sinisi R, Tommasi S, Umani-Ronchi A (2006) J Am Chem Soc 128:1424–1425
13. Kuroda N, Takahashi Y, Yoshinaga K, Mukai C (2006) Org Lett 8:1843–1845
14. Yasuhara A, Sakamoto T (1998) Tetrahedron Lett 39:595–596
15. Lombardino JG, Wiesman EH (1971) J Med Chem 14:973–977
16. Lombardino JG, Wiesman EH, McLamore WM (1971) J Med Chem 14:1171–1175
17. Lombardino JG, Wiesman EH (1972) J Med Chem 15:848–849
18. Zinnes H, Lindo NA, Sircar JC, Schwartz ML, Shavel J Jr (1973) J Med Chem 16:44–48
19. Zinnes H, Sircar JC, Lindo N, Schwartz ML, Fabian AC, Shavel J Jr, Kasulanis CF, Genzer JD, Lutomski C, DiPasquale G (1982) J Med Chem 25:12–18
20. Kwon S-K, Park M-S (1992) Arch Pharm Res 15:251–255

21. Lazer ES, Miao CK, Cywin CL, Sorcek R, Wong H-C, Meng Z, Potocki I, Hoermann M, Snow RJ, Tschantz MA, Kelly TA, McNeil DW, Coutts SJ, Churchill L, Graham AG, David E, Grob PM, Engel W, Meier H, Trummlitz G (1997) J Med Chem 40:980–989
22. Lee EB, Kwon SK, Kim SG (1999) Arch Pharm Res 22:44–47
23. Watanabe H, Mao C-L, Barnish IT, Hauser CR (1969) J Org Chem 34:919–926
24. Lombardino JG, Kuhla DE (1981) Adv Heterocycl Chem 28:73–126
25. Motherwell WB, Pennell AMK (1991) J Chem Soc Chem Commun 877–879
26. Nemazanyi AG, Volovenko YM, Neshchadimenko VV, Babichev FS (1992) Chem Heterocycl Comp 28:220–222
27. Manjarrez N, Pérez HI, Sorís A, Luna H (1996) Synth Commun 26:585–591
28. Manjarrez N, Pérez HI, Sorís A, Luna H (1996) Synth Commun 26:1405–1410
29. Takahashi M, Morimoto T, Isogai K, Tsuchiya S, Mizumoto K (2001) Heterocycles 55:1759–1769
30. Layman WJ, Greenwood TD, Downey AL, Wolfe JF (2005) J Org Chem 70:9147–9155
31. Vidal A, Madelmont J-C, Mounetou E (2006) Synthesis 591–593
32. Aliyenne AO, Kraïem J, Kacem Y, Hassine BB (2008) Tetrahedron Lett 49:1473–1475
33. Zia-ur-Rehman M, Choudary JA, Elsegood MRJ, Siddiqui HL, Khan KM (2009) Eur J Med Chem 44:1311–1316
34. Barange DK, Batchu VR, Gorja D, Pattabiraman VR, Tatini LK, Babu JM, Pal M (2007) Tetrahedron 63:1775–1789
35. Barange DK, Nishad TC, Swamy NK, Bandameedi V, Kumar D, Sreekanth BR, Vyas K, Pal M (2007) J Org Chem 72:8547–8550
36. Hatano M, Mikami K (2003) J Am Chem Soc 125:4704–4705
37. Bressy C, Alberico D, Lautens M (2005) J Am Chem Soc 127:13148–13149
38. Marchal E, Uriac P, Legouin B, Toupet L, van de Weghe P (2007) Tetrahedron 63:9979–9990
39. Parmentier J-G, Poissonnet G, Goldstein S (2002) Heterocycles 57:465–476
40. Costa M, Cá ND, Gabriele B, Massera C, Salerno G, Soliani M (2004) J Org Chem 69:2469–2477
41. Sakai S, Annnaka K, Konakahara T (2006) J Org Chem 71:3653–3655
42. Arcadi A, Bianchi G, Marinelli F (2004) Synthesis 610–618
43. Vlasov VM, Terekhova MI, Petrov ES, Sutula VD, Shatenshtein AI (1982) Zhurnal Organicheskoi Khimii 18:1672–1679
44. Larock RC, Fried CA (1990) J Am Chem Soc 112:5882–5884

Chapter 3
Facile Synthesis of 1,2,3,4-Tetrahydro-β-Carbolines by One-Pot Domino Three-Component Indole Formation and Nucleophilic Cyclization

A 1,2,3,4-tetrahydro-β-carboline, which consists of a tricyclic indole, is an attractive drug template due to its potential antioxidative activity [1–7]. Carboline derivatives are also useful as intermediates for natural product synthesis [8–23]. Because construction of tetrahydro-β-carbolines is mostly dependent on the Pictet-Spengler [8–17] and related reactions [18–23], development of alternative synthetic methodologies is extremely important to ensure diversity-oriented synthesis. For other representative synthetic routes, see: [24–31].

In Chap. 1, the author reported the copper-catalyzed synthesis of 2-(aminomethyl)indoles via a domino three-component coupling-cyclization reaction of a 2-ethynylanilines, paraformaldehyde and a secondary amine [32, 33]. For related heterocycle syntheses, see [34, 35]. Bosch and co-workers [36] previously reported that treatment of a 2-[N-(benzenesulfonyl)indol-2-yl]piperidin-4-one derivative having an N-hydroxylethyl group with t-BuOK brought about the formation of the corresponding indolo[2,3-a]quinolizine, although this was an isolated example. On the other hand, it is well established that cyclization at the 3-position of N-alkylindoles containing an ester group is efficiently promoted by a strong acid to afford 4-oxo-tetrahydro-β-carbolines [37–42]. Based on these chemistries, the author expected that 2-(aminomethyl)indole 5, generated by copper-catalyzed indole formation using ethynylanilines 1, aldehydes 2, and secondary amines 3 bearing an appropriate functionality (R^3 = CH_2OH or CO_2R), could be converted into β-carboline derivatives 6 or 7 by a second cyclization at the C-3 position (Scheme 1). This sequential reaction is challenging in that various reactive components exist in the reaction mixture, including unprotected amine(s), an aldehyde, and an ester/alcohol, especially when N-alkylanilines are employed. In this Section, the author reports two direct routes to 1,2,3,4-tetrahydro-β-carboline derivatives by a copper-catalyzed three-component coupling-indole formation–nucleophilic cyclization at the 3-position. To the best of the author's knowledge, there is no precedent for multi-component synthesis of tetrahydro-β-carbolines, except for those using the Pictet-Spengler type reaction [8–17, 43, 44].

The initial attempt was carried out with N-tosyl-2-ethynylaniline **1a**, butanal **2a** (2 equiv.), and 2-(N-methylamino)ethanol **3a** (1.1 equiv.) in the presence of 5 mol

Y. Ohta, *Copper-Catalyzed Multi-Component Reactions*, Springer Theses,
DOI: 10.1007/978-3-642-15473-7_3, © Springer-Verlag Berlin Heidelberg 2011

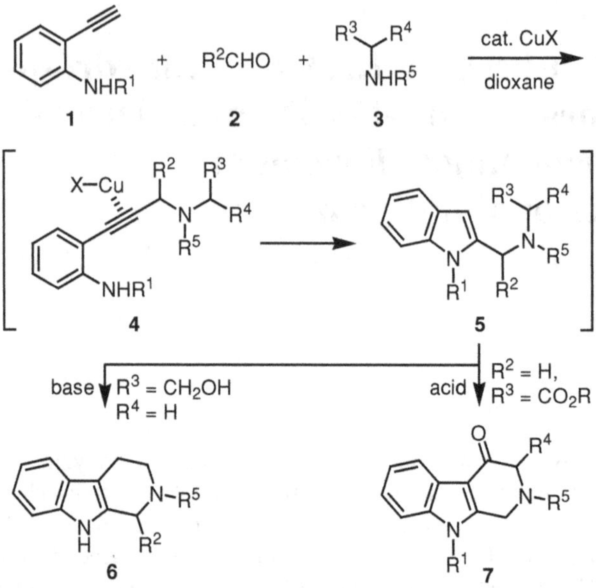

Scheme 1 Two direct routes to 1,2,3,4-tetrahydro-β-carboline derivatives

% CuBr (Table 1). After the three-component indole formation in dioxane was completed (monitored by TLC, 80 °C for 1 h), t-BuOK (3 equiv.) was added to the reaction mixture. Although the desired bis-cyclization product 1,2,3,4-tetrahydro-β-carboline derivative **6a** was obtained in 31% yield, the N-cyclization product **8a** was formed as the major product (69% yield, entry 1). The author has already reported a selective N-cyclization with an aryl bromide moiety, see: [35]. To improve the selectivity of the second cyclization, the author optimized the reaction conditions for deprotection–cyclization as well as the nitrogen protecting group. Bosch proposed that the arylsulfonyl group on the indole nitrogen would be transferred to the primary hydroxy group by the action of in situ-generated t-BuOTs, and nucleophilic attack of the C-3 position of the resulting NH-indole furnishes the corresponding cyclization product, see [36]. Addition of Et$_2$O as the co-solvent slightly improved the selectivity but decreased the combined yield to 43% (entry 2). In contrast, use of hexane led to the formation of **6a** as the major product (53% yield, entry 3). These results are in good agreement with Bosch's observation, in which carrying out the reaction in a less polar solvent improved the selectivity of the C-3 cyclization over the N-cyclization [36]. As the N-protecting group of 2-ethynylaniline, mesyl and mesitylenesulfonyl (Mts) groups were less effective for selective formation of **6a** (entries 4 and 5). The reaction of **1d** bearing an N-benzenesulfonyl group gave a better result (entry 6) than that of the N-tosyl derivative **1a** (entry 3). This result promoted the author to utilize more electron deficient benzenesulfonamides **1e–h** bearing a halogen atom or nitro group on the benzene ring (entries 7–10). The results indicated that 4-chlorophenylsulfonyl group was the best protecting group of the aniline nitrogen (entry 8). In this case,

Table 1 One-pot three-component synthesis of tetrahydro-β-carbolines using t-BuOK

Entry	R^1	Conditions	Co-solvent	Yield (%)^a	
				6a	8a
1	Ts (**1a**)	80 °C, 1 h	–	31	69
2	Ts (**1a**)	80 °C, 1 h	Et$_2$O	23	20
3	Ts (**1a**)	80 °C, 1 h	Hexane	53	33
4	Ms (**1b**)	80 °C, 2 h	Hexane	29	35
5	Mts (**1c**)	80 °C, 2 h	Hexane	19	43
6	SO$_2$Ph (**1d**)	80 °C, 1.5 h	Hexane	63	25
7	SO$_2$C$_6$H$_4$(4-Br) (**1e**)	80 °C, 0.5 h	Hexane	58	14
8	SO$_2$C$_6$H$_4$(4-Cl) (**1f**)	80 °C, 0.5 h	Hexane	65	18
9	SO$_2$C$_6$H$_4$(4-F) (**1g**)	80 °C, 0.5 h	Hexane	48	20
10	SO$_2$C$_6$H$_4$(4-NO$_2$) (**1h**)	80 °C, 0.5 h	Hexane	23	10
11	SO$_2$C$_6$H$_4$(4-Cl) (**1f**)	50 °C, 1.5 h	Hexane	75	25

Ethynylaniline **1** (0.18 mmol), n-PrCHO **2a** (2 equiv.), and 2-(N-methylamino)ethanol **3a** (1.1 equiv.) in dioxane (2 mL) were treated with CuBr (5 mol %) under the conditions shown in the table. After the indole formation was completed (monitored by TLC), co-solvent (2 mL) and t-BuOK (3 equiv.) were added at 0 °C and the reaction mixture was stirred at 0 °C for 5 min and rt for an additional 30 min
^{a} Isolated yields

the 2,3-unsubstituted N-arylsulfonylindoles, formed by intramolecular hydroamination of **1** without resulting in a Mannich-type reaction, were observed as a byproduct. The author also tested the reaction at 50 °C for the three-component indole formation and obtained **6a** in 75% yield (entry 11). When NaH or KH was used instead of t-BuOK, the desired product **6a** was not obtained. This suggest that the C-3 cyclization proceeds through rearrangement of the arylsulfonyl group from the nitrogen atom of the indole to the hydroxyl group, as proposed by Bosch et al.

Under the optimized conditions (Table 1, entry 11), the scope of this one-pot tetrahydro-β-carboline synthesis was explored using ethynylaniline derivative **1f** and several aldehydes (Table 2). Reaction with aldehyde **2b** or **2c** containing a (trimethylsilyl)vinyl or benzyloxymethyl group afforded **6b** and **6c** in moderate yields (entries 1 and 2, 48 and 55%, respectively), accompanied by the by-products **8b** and **8c**, respectively. In these reactions, a prolonged reaction time and elevated temperature were necessary for completion of the initial indole formation,

Table 2 Synthesis of tetrahydro-β-carbolines using several aldehydes

Entry	Aldehyde	Conditions[b]	Product (yield)[c,d]
1	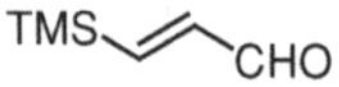 TMS~~~CHO **2b**	50 °C, 2 h then 100 °C, 0.5 h	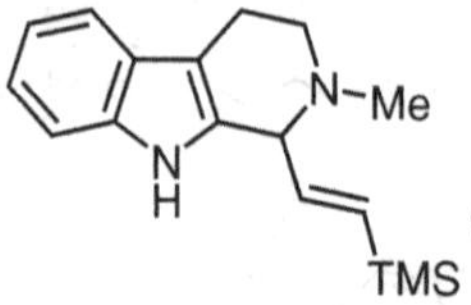 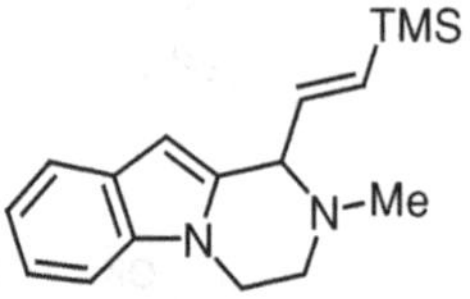 **6b** (48%)　　　**8b** (8%)
2	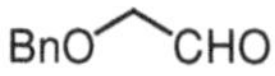 BnO⌒CHO **2c**	50 °C, 2 h then 100 °C, 0.5 h	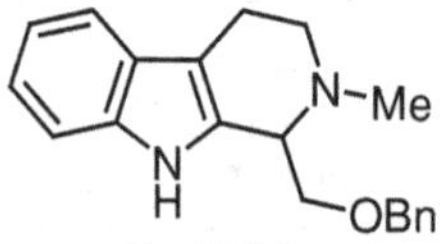 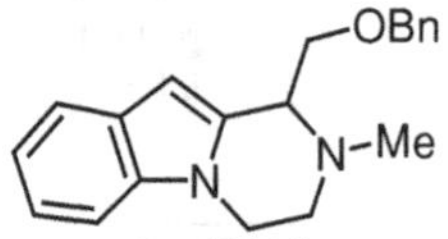 **6c** (55%)　　　**8c** (16%)
3	(HCHO)$_n$ **2d**	50 °C, 0.5 h	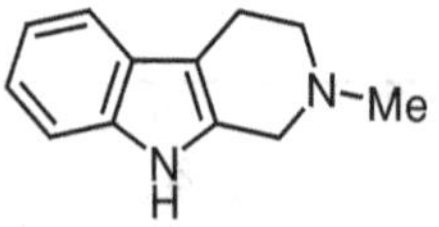 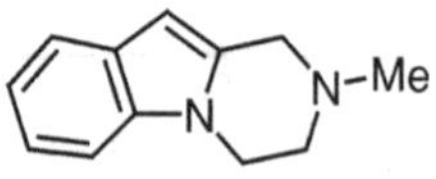 **6d** (45%)　　　**8d** (0%)[e]

Ethynylaniline **1f** (0.18 mmol), aldehyde **2** (2 equiv.), and 2-(N-methylamino)ethanol **3a** in dioxane were treated with CuBr (5 mol %) under the conditions shown in the table. Then hexane (2 mL) and t-BuOK were added at 0 °C and the reaction mixture was stirred at 0 °C for 5 min and rt for an additional 30 min

[a] Conditions for the initial indole formation
[b] Isolated yields
[c] Structures of **8b–d** are shown below
[d] Not isolated

presumably because of the steric bulkiness of the functional groups. The reaction with paraformaldehyde **2d** gave **6d** in 45% yield (entry 3). The high polarity of **6d** considerably lowered the chemical yield during purification with column chromatography over silica gel. Use of alumina column partly improved the yield of **6d** (45%).

The author next investigated the acid-induced direct construction of a 4-oxo-tetrahydro-β-carboline scaffold using amino esters **3b–j** (Table 3). In this reaction, use of anilines without an electron-withdrawing group on the nitrogen atom is essential to secure the nucleophilicity of the intermediate indoles of type **5** (Scheme 1). A mixture of N-methyl-2-ethynylaniline **1i**, paraformaldehyde **2d**, and N-methylglycine ethyl ester **3b** was treated with 5 mol % of CuBr in dioxane at 170 °C under microwave irradiation (condition A) followed by the reaction with MsOH. Other acids were less effective. For example, after indole formation with **1g**, **2d**, and **3d** was completed, the reaction mixture was treated with polyphosphoric acid (PPA) to give **7c** in only 19% yield to give the desired 4-oxo-1,2,3,4-tetrahydro-β-carboline **7a** in 72% yield (entry 1). The N-allyl or N-butylglycine derivatives **3c** and **3d** showed clean conversion to **7b** and **7c**, respectively (entries 2 and 3). Methyl ester **3e** was also a good component for this one-pot reaction (entry 4). Whereas **3f** having an N-benzyl group resulted in sluggish conversion in

Table 3 Preparation of 4-oxo-tetrahydro-β-carboline by domino three-component coupling-indole formation and successive MsOH-induced cyclization

Entry	Amino esters	Conditions[b]	Product (yield)[c]
	RHN⌒CO₂Et	A	7a / 7b
1	**3b**: R = Me		**7a** (72%)
2	**3c**: R = allyl		**7b** (77%)
	BuHN⌒CO₂R		
3	**3d**: R = Et	A	**7c** (70%)
4	**3e**: R = Me	A	**7c** (68%)
	BnHN⌒CO₂Me		
5	**3f**	A	**7d** (32%)
6	**3f**	B	**7d** (57%)
	MeHN–CO₂Me (R)		
7	**3g**: R = Me	C	**7e**: R = Me (63%)
8	**3h**: R = i-Bu	C	**7f**: R = i-Bu (37%)
9	**3i**: R = Bn	C	**7g**: R = Bn (46%)
10	**3j**	C	**7h** (29%)

The mixture of ethynylaniline **1i** (0.19 mmol), paraformaldehyde **2d** (2 equiv.), and amino ester **3** (1.2 equiv.) in dioxane was stirred with CuX (5 mol%) under microwave irradiation (300 W). After indole formation was complete on TLC, the reaction mixture was treated with MsOH at 80 °C for 30 min

[a] Condition A: CuI, 170 °C, 1 h; condition B: CuBr, 120 °C, 15 min, then 140 °C, 15 min; condition C: CuBr, 120 °C, 15 min

[b] Isolated yields

the indole formation step using condition A (entry 5), use of CuBr, a more reactive catalyst for the initial three-component indole formation than CuI, led to 57% yield of **7d** after treatment with MsOH (condition B, entry 6). This one-pot construction of β-carboline derivatives also tolerated such chiral amino acid derivatives as **3g–i** (entries 7–9). The tetracyclic compound **7h** can be easily obtained from racemic pipecolinate **3j**, although in relatively low yield (29%, entry 10). It should be noted that the indole formation of Mannich adducts derived from **1i** did not proceed when using aldehydes other than paraformaldehyde and amino esters.

In conclusion, the author has developed two direct synthetic routes to 1,2,3,4-tetrahydro-β-carboline derivatives by copper-catalyzed three-component indole formation followed by successive cyclization at the 3-position of indole. When an aminoethanol was used as the amine component, the 4-chlorophenylsulfonyl group is the protecting/activating group of choice for the second cyclization induced by *t*-BuOK. On the other hand, *N*-methyl-2-ethynylaniline and α-amino esters were good components for MsOH-induced cyclization at C-3 to produce various 4-oxo-1,2,3,4-tetrahydro-β-carbolines, including optically active ones. These two methodologies using three-component coupling of readily available substrates should contribute to diversity-oriented synthesis of tetrahydro-β-carbolines as a drug-like scaffold.

3.1 Experimental Section

The compounds **2a** and **2c** are commercially available.

The compounds **1a**, **1b** [45], **1d** [46], **1h** [47], **1i** [48], **2b** [49], **3b** [50], **3c** [51], **3d** [52], **3e** [53], **3f**, **3g** [54], **3h** [55], **3i** [56], **3j**[57] are known.

3.1.1 General Methods

IR spectra were determined on a JASCO FT/IR-4100 spectrometer. Exact mass (HRMS) spectra were recorded on JMS-HX/HX 110A mass spectrometer. ^{1}H NMR spectra were recorded using a JEOL AL-500 spectrometer at 500 MHz frequency. Chemical shifts are reported in δ (ppm) relative to Me$_4$Si (in CDCl$_3$) as internal standard. ^{13}C NMR spectra were recorded using a JEOL AL-500 and referenced to the residual CHCl$_3$ signal. Optical rotations were measured with a JASCO P-1020 polarimeter. Melting points were measured by a hot stage melting points apparatus (uncorrected). Microwave reaction was conducted in a sealed glass vessel (capacity 10 mL) using CEM Discover microwave reactor with a run time of no more than 10 min. The temperature was monitored using IR sensor mounted under the reaction vessel. For column chromatography, Wakosil C-300 was employed. For HPLC separations, a CHIRALCEL OD-H analytical column (DICEL CHEMICAL INDUSTRIES LTD., 4.6 × 150 mm, flow rate 0.5 mL/min) was employed, and eluting products were detected by UV at 256 nm. A solvent system consisting of 0.1% Et$_2$NH in n-hexane (v/v, solvent A) and 0.1% Et$_2$NH in i-PrOH (v/v, solvent B) was used for HPLC elution with a linear gradient of i-PrOH (20–40% over 45 min).

3.1.2 General Procedure for Synthesis of N-Arylsulfonyl-2-ethynylaniline: Synthesis of 2-Ethynyl-N-mesitylenesulfonylaniline (1c)

To a stirred solution of 2-ethynylaniline (0.30 g, 2.56 mmol), pyridine (1.04 mL, 12.80 mmol), and DMAP (6 mg, 0.05 mmol) in CH$_2$Cl$_2$ (15 mL) was added Mts-Cl (0.67 g, 3.07 mmol) at 0 °C under Ar. The mixture was stirred at rt for 12 h and washed with 2 N HCl, H$_2$O and brine. The Organic layer was dried over MgSO$_4$ and concentrated under reduced pressure. The residue was purified by column chromatography over silica gel with hex/EtOAc (10:1) as the eluent to give 1c (0.76 g, quant.) as a colorless solid which was recrystallized from hex-AcOEt as colorless crystals: mp 96–98 °C; IR (neat) 2,103 cm^{-1} (C≡C); ^{1}H NMR (500 MHz, CDCl$_3$) δ 2.26 (s, 3H, CH$_3$), 2.68 (s, 6H, 2 × CH$_3$), 3.44 (s, 1H, C≡C), 6.92 (s, 2H, Ar), 6.96 (dd, J = 7.7, 7.7 Hz, 1H, Ar), 7.20 (ddt J = 7.7, 7.7, 1.4 Hz, 1H, Ar), 7.29 (d, J = 7.7 Hz, 1H, Ar), 7.38 (dd, J = 7.7, 1.4 Hz, 1H, Ar), 7.41 (brs, 1H, NH); ^{13}C NMR (125 MHz, CDCl$_3$) δ 20.1, 23.1 (2C), 78.8, 84.4, 111.3, 117.1, 123.3, 130.1, 132.2, 132.7, 133.3, 138.8, 139.5 (2C), 142.9. Anal. Calcd. for C$_{17}$H$_{17}$NO$_2$S: C, 68.20; H, 5.72; N, 4.68. Found C, 68.09; H, 5.81; N, 4.66.

3.1.3 N-(p-Bromobenzenesulfonyl)-2-ethynylaniline (1e)

To a stirred solution of 2-ethynylaniline (0.20 g, 1.71 mmol) in pyridine (10 mL) was added p-bromobenzenesulfonyl chloride (0.52 g, 2.05 mmol) at 0 °C under Ar. The mixture was stirred for 12 h at rt and the washed with 2 N HCl, H$_2$O, and brine. The Organic layer was dried over MgSO$_4$ and concentrated under reduced pressure. The residue was purified by column chromatography over silica gel with hex/EtOAc (10:1) as the eluent to give 1c (0.57 g, quant.) as a colorless solid which was recrystallized from hex-AcOEt as colorless crystals: mp 62–64 °C; IR (neat) 2,105 cm^{-1} (C≡C); ^{1}H NMR (500 MHz, CDCl$_3$) δ 3.36 (s, 1H, C≡C), 7.05–7.08 (m, 1H, Ar), 7.20 (brs, 1H, NH), 7.30–7.34 (m, 1H, Ar), 7.36 (dd, J = 7.7, 1.4 Hz, 1H, Ar), 7.55–7.57 (m, 2H, Ar), 7.59 (d, J = 9.2 Hz, 1H, Ar), 7.63–7.66 (m, 2H, Ar); ^{13}C NMR (125 MHz, CDCl$_3$) δ 78.5, 84.5, 113.2, 120.0, 124.8, 128.3, 128.8 (2C), 130.3, 132.3 (2C), 132.7, 137.85, 137.90. Anal. Calcd. for C$_{14}$H$_{10}$BrNO$_2$S: C, 50.01; H, 3.00; N, 4.17. Found C, 50.01; H, 3.14; N, 4.10.

3.1.4 N-(p-*Chlorobenzenesulfonyl*)-2-*ethynylaniline (1f)*

By a procedure similar to that described for **1e**, 2-ethynylaniline (0.20 g, 1.71 mmol) was converted into **1f** (0.50 g, quant.) by treatment with *p*-chloro-benzenesulfonyl chloride (0.43, 2.05 mmol); colorless crystals (from CHCl₃–hexane): mp 69–70 °C; IR (neat) 2,109 cm⁻¹ (C≡C); ¹H NMR (500 MHz, CDCl₃) δ 3.35 (s, 1H, C≡C), 7.06 (dd, J = 7.7 Hz, 1H, Ar), 7.20 (brs, 1H, NH), 7.30–7.34 (m, 1H, Ar), 7.36 (dd, J = 7.7, 1.4 Hz, 1H, Ar), 7.38–7.41 (m, 2H, Ar), 7.60 (d, J = 7.7 Hz, 1H, Ar), 7.70–7.73 (m, 2H, Ar); ¹³C NMR (125 MHz, CDCl₃) δ 77.3, 84.5, 113.2, 120.0, 124.8, 128.8 (2C), 129.3 (2C), 130.3, 132.7, 137.3, 137.9, 139.8. *Anal.* Calcd. for C₁₄H₁₀ClNO₂S: C, 57.63; H, 3.45; N, 4.80. Found C, 57.35; H, 3.60; N, 4.80.

3.1.5 2-*Ethynyl*-N-(p-*fluorosulfonyl*)*aniline (1g)*

By a procedure similar to that described for **1e**, 2-ethynylaniline (0.50 g, 4.26 mmol) was converted into **1g** (1.17 g, quant.) by treatment with *p*-fluoro-benzenesufonyl chloride (1.19, 6.13 mmol); colorless crystals (from CHCl₃–hexane): mp 74–75 °C; IR (neat) 2,104 cm⁻¹ (C≡C); ¹H NMR (500 MHz, CDCl₃) δ 3.34 (s, 1H, C≡C), 7.04–7.07 (m, 1H, Ar), 7.08–7.11 (m, 2H, Ar), 7.18 (brs, 1H, NH), 7.30–7.34 (m, 1H, Ar), 7.35 (dd, J = 7.7, 1.4 Hz, 1H, Ar), 7.60 (d, J = 9.2 Hz, 1H, Ar), 7.77–7.81 (m, 7H, Ar); ¹³C NMR (125 MHz, CDCl₃) δ 78.5, 84.4, 113.3, 116.3 (d, J = 22.8 Hz, 2C), 120.1, 124.7, 130.1 (d, J = 9.6 Hz, 2C), 130.3, 132.6, 134.8, 138.0, 165.4 (d, J = 257.9 Hz). *Anal.* Calcd. for C₁₄H₁₀FNO₂S: C, 61.08; H, 3.66; N, 5.09. Found C, 60.80; H, 3.78; N, 5.00.

3.1.6 *General Procedure for Synthesis of 1,2,3,4-Tetrahydro-β-carboline by Domino Copper-Catalyzed Three-Component Indole Formation and Cyclization with t-BuOK: Synthesis of 2-Methyl-1-propyl-2,3,4,9-tetrahydro-1H-pyrido[3,4-b]indole (6a) and 2-Methyl-1-propyl-1,2,3,4-tetrahydropy-razino[1,2-a]indole (8a) (Table 1, Entry 11)*

A mixture of *N*-(4-chlorophenyl)sulfonyl-2-ethynylaniline **1f** (53.6 mg, 0.18 mmol), butanal **2a** (33.2 µL, 0.37 mmol), 2-(*N*-methylamino)ethanol **3a** (16.3 µL, 0.21 mmol), and CuBr (1.3 mg, 0.0092 mmol) in dioxane (1 mL) was stirred at 50 °C for 1.5 h [for the reaction with **2b** (Table 2, entry 1) and **2c** (Table 2, entry 2), the mixture was stirred at 100 °C for an additional 0.5 h]. After the three-component indole formation was completed on TLC, hexane (2 mL) was added at rt and the mixture was cooled to 0 °C. *t*-BuOK (62.0 mg, 0.55 mmol) was added at

0 °C and the reaction mixture was stirred for 5 min at 0 °C and additional 30 min at rt. The reaction mixture was concentrated under reduced pressure and purified by column chromatography over silica gel with hexane/EtOAc (3:1 to 1:3) as the eluent to give **6a** (31.8 mg, 75%) and **8a** (10.7 mg, 25%) both as an yellow oil.

Compound **6a**:^{1}H NMR (500 MHz, CDCl$_3$) δ 0.94 (t, $J = 7.2$ Hz, 3H, CH$_2$C*H$_3$*), 1.31–1.41 (m, 1H, C*H*H), 1.45–1.56 (m, 1H, CH*H*), 1.69–1.76 (m, 1H, C*H*H), 1.81–1.89 (m, 1H, CH*H*), 2.47 (s, 3H, NMe), 2.69–2.82 (m, 3H, 3 × CH), 3.14–3.20 (m, 1H, CH), 3.51 (t, $J = 5.4$ Hz, 1H, 1-H), 7.08–7.11 (m, 1H, Ar), 7.12–7.15 (m, 1H, Ar), 7.31 (d, $J = 8.0$ Hz, 1H, Ar), 7.48 (d, $J = 7.4$ Hz, 1H, Ar), 7.72 (brs, 1H, NH); ^{13}C NMR (125 MHz, CDCl$_3$) δ 14.3, 18.7, 19.0, 35.3, 41.9, 49.6, 59.8, 108.2, 110.6, 118.0, 119.3, 121.3, 127.3, 135.1, 135.8; MS (FAB) m/z (%): 229 (MH$^+$, 50), 185 (100); HRMS (FAB) calcd for C$_{15}$H$_{21}$N$_2$ (MH$^+$): 229.1705; found: 229.1713.

Compound **8a**: ^{1}H NMR (500 MHz, CDCl$_3$) δ 0.94 (t, $J = 7.4$ Hz, 3H, CH$_2$C*H$_3$*), 1.34–1.53 (m, 2H, C*H$_2$*CH$_3$), 1.82–1.89 (m, 1H, C*H*H), 1.91–1.99 (m, 1H, CH*H*), 2.43 (s, 3H, NMe), 2.92 (ddd, $J = 12.6, 9.2, 4.6$ Hz, 1H, C*H*H), 3.26–3.30 (m, 1H, CH*H*), 3.65 (t, $J = 4.6$ Hz, 1H, 1-H), 4.01–4.10 (m, 2H, CH$_2$), 6.24 (s, 1H, 9-H), 7.08–7.11 (m, 1H, Ar), 7.13–7.17 (m, 1H, Ar), 7.26 (d, $J = 8.0$ Hz, 1H, Ar), 7.56 (d, $J = 7.4$ Hz, 1H, Ar); ^{13}C NMR (125 MHz, CDCl$_3$) δ 14.4, 17.8, 34.2, 39.9, 41.0, 50.9, 60.8, 96.8, 108.6, 119.7, 119.9, 120.5, 128.2, 136.0, 138.1; MS (FAB) m/z (%): 229 (MH$^+$, 50), 185 (100); HRMS (FAB) calcd for C$_{15}$H$_{21}$N$_2$ (MH$^+$): 229.1705; found: 229.1703.

3.1.7 2-Methyl-1-[2-(trimethylsilyl)ethenyl]-2,3,4,9-tetrahydro-1H-pyrido[3,4-b]indole (6b) and 2-Methyl-1-[2(trimethyl-silyl)ethenyl]-1,2,3,4-tetrahydropyrazino[1,2-a]indole (8b)

By a procedure similar to that described for indole **6a** and **8a**, **1f** (53.6 mg, 0.18 mmol) was converted into **6b** (25.0 mg, 48%) and **8b** (4.2 mg, 8%) both as an yellow oil by treatment with (*E*)-3-(trimethylsilyl)acrylaldehyde **2b** (47.2 mg, 0.37 mmol).

Compound **6b**: ^{1}H NMR (500 MHz, CDCl$_3$) δ 0.13 (s, 9H, SiMe$_3$), 2.47 (s, 3H, Me), 2.60–2.66 (m, 1H, CH*H*), 2.75–2.80 (m, 1H, C*H*H), 2.90–2.97 (m, 1H, CH*H*), 3.12–3.18 (m, 1H, C*H*H), 3.81 (d, 1H, $J = 8.0$ Hz, CH), 6.00 (dd, 1H, $J = 18.3, 8.0$ Hz, CHC*H*CH), 6.10 (d, 1H, C*H*SiMe$_3$ $J = 18.3$ Hz), 7.08–7.91 (m, 1H, Ar), 7.15 (ddd, 1H, $J = 7.4, 7.4, 1.1$ Hz, Ar), 7.31 (d, 1H, $J = 8.0$ Hz, Ar), 7.50 (d, 1H, $J = 7.4$ Hz, Ar), 7.55 (br, 1H, NH); ^{13}C NMR (125 MHz, CDCl$_3$) δ −1.20 (3C), 21.3, 43.4, 52.2, 68.4, 108.3, 110.8, 118.3, 119.3, 121.5, 127.5, 132.7, 135.8, 136.2, 145.7; MS (FAB) m/z (%): 73 (100), 185 (88), 285 (MH$^+$, 47); HRMS (FAB) calcd for C$_{17}$H$_{25}$N$_2$Si (MH$^+$): 285.1787; found: 285.1794.

Compound **8b**: ^{1}H NMR (500 MHz, CDCl$_3$) δ 0.14 (s, 9H, SiMe$_3$), 2.44 (s, 3H, Me), 2.76–2.82 (m, 1H, CH*H*), 3.24 (ddd, 1H, $J = 12.0, 6.0, 3.0$ Hz, C*H*H), 3.84 (d, 1H, $J = 6.9$ Hz, CH), 4.03–4.09 (m, 1H, CH*H*), 4.17 (d, 1H, $J = 11.5, 5.7$,

3.0 Hz, CH*H*), 5.99 (dd, 1H, *J* = 18.9, 6.9 Hz, CHC*H*CH), 6.05 (d, 1H, *J* = 18.9 Hz, C*H*SiMe₃), 6.09 (s, 1H, Ar), 7.08–7.11 (m, 1H, Ar), 7.14–7.18 (m, 1H, Ar), 7.28 (d, 1H, *J* = 8.0 Hz, Ar), 7.55 (d, 1H, *J* = 7.4 Hz, Ar); ^{13}C NMR (125 MHz, CDCl₃) δ −1.26 (3C), 41.6, 43.6, 51.9, 69.3, 98.4, 108.7, 119.8, 120.2, 120.7, 128.1, 135.2, 136.3, 136.4, 114.4; MS (FAB) *m/z* (%):, 185 (50), 285 (MH⁺, 25); HRMS (FAB) calcd for C₁₇H₂₅N₂Si (MH⁺): 285.1787; found: 285.1805.

3.1.8 1-(Benzyloxymethyl)-2-methyl-2,3,4,9-tetrahydro-1H-pyr-ido[3,4-b]indole (6c) and 1-(Benzyloxymethyl)2-methyl-1,2,3,4-tetrahydropyrazino[1,2-a]indole (8c)

By a procedure similar to that described for indole **6a** and **8a**, **1f** (53.6 mg, 0.18 mmol) was converted into **6c** (31.7 mg, 55%) and **8c** (9.1 mg, 16%) both as an yellow oil by treatment with benzyloxyacetaldehyde **2c** (51.9 μL, 0.37 mmol).

Compound **6c**: ^{1}H NMR (500 MHz, CDCl₃) δ 2.54 (s, 3H, Me), 2.75–2.83 (m, 3H, 3 × CH*H*), 3.08–3.11 (m, 1H, CH*H*), 3.57 (dd, 1H, *J* = 9.2, 9.2 Hz, CH), 3.69 (dd, 1H, *J* = 9.2, 4.0 Hz, CHCH*H*), 4.01 (dd, 1H, *J* = 9.2, 4.0 Hz, CHCH*H*), 4.61 (d, 1H, *J* = 12.2 Hz, PhCH*H*), 4.63 (d, 1H, *J* = 12.2 Hz, PhCH*H*), 7.07–7.10 (m, 1H, Ar), 7.13–7.15 (m, 1H, Ar), 7.28 (d, 1H, *J* = 8.0 Hz, Ar), 7.31–7.41 (m, 5H, Ar), 7.50 (d, 1H, *J* = 7.4 Hz, Ar), 8.48 (br, 1H). ^{13}C NMR (125 MHz, CDCl₃) δ 20.0, 43.4, 52.1, 59.5, 72.8, 73.8, 107.8, 110.8, 118.1, 119.0, 121.3, 126.5, 127.9 (2C), 128.0, 128.6 (2C), 134.4, 135.8, 137.7; MS (FAB) *m/z* (%): 185 (100), 307 (MH⁺, 45); HRMS (FAB) calcd for C₂₀H₂₃N₂O (MH⁺): 307.1810; found: 307.1813.

Compound **8c**: ^{1}H NMR (500 MHz, CDCl₃) δ 2.57 (s, 3H, Me), 2.93–2.98 (m, 1H, CH*H*), 3.28–3.32 (m, 1H, CH*H*), 3.81–3.90 (m, 3H, CH, 2 × CH*H*), 4.04–4.14 (m, 2H, 2 × CH*H*), 4.59 (d, 1H, *J* = 12.0 Hz, PhCH*H*), 4.63 (d, 1H, *J* = 12.0 Hz, PhCH*H*), 6.26 (s, 1H, Ar), 7.08–7.11 (m, 1H, Ar), 7.15–7.18 (m, 1H, Ar), 7.27–7.33 (m, 6H, Ar), 7.55 (d, 1H, *J* = 7.4 Hz, Ar); ^{13}C NMR (125 MHz, CDCl₃) δ 39.9, 42.4, 50.8, 61.0, 71.8, 73.5, 97.4, 108.6, 109.8, 120.11, 120.70, 127.7, 127.9 (2C), 128.0, 128.4 (2C), 135.1, 135.9, 138.0; MS (FAB) *m/z* (%): 185 (100), 307 (MH⁺, 35); HRMS (FAB) calcd for C₂₀H₂₃N₂O (MH⁺): 307.1810; found: 307.1810.

3.1.9 2-Methyl-2,3,4,9-tetrahydro-1H-pyrido[3,4-b]indole (6d)

By a procedure similar to that described for indole **6a** and **8a**, **1f** (53.6 mg, 0.18 mmol) was converted into **6d** (17.2 mg, 45%) by treatment with (HCHO)ₙ and **2d** (12.4 mg, 0.37 mmol); colorless crystals: mp (from CHCl₃–hexane): 212 °C; ^{1}H NMR (500 MHz, CDCl₃) δ 2.49 (s, 3H, NMe), 2.76–2.84 (m, 4H, CH₂CH₂), 3.57 (s, 2H, ArCH₂), 7.06–7.09 (m, 1H, Ar), 7.11–7.14 (m, 1H, Ar),

7.27 (d, 1H, $J = 7.4$ Hz, Ar), 7.47 (d, 1H, $J = 7.4$ Hz, Ar), 7.91 (br, 1H, NH); ^{13}C NMR (125 MHz, CDCl$_3$) δ 21.5, 45.3, 51.8, 53.0, 107.7, 110.9, 117.9, 119.1, 121.2, 127.1, 131.9, 136.1. Anal. calcd. for C$_{12}$H$_{14}$N$_2$: C, 77.38; H, 7.58; N, 15.04. Found C, 77.38; H, 7.58; N, 15.04.

3.1.10 General Procedure for Synthesis of 1,2,3,4-Tetrahydro-β-carboline by Domino Copper-Catalyzed Three-Component Indole Formation and Cyclization with MsOH: Synthesis of 2-Methyl-2,3-dihydropyrido[3,4-b]indol-4(9H)-one (7a) (Conditions A)

A mixture of N-methyl-2-ethynylaniline 1i (25.0 mg, 0.19 mmol), paraformaldehyde 2d (11.4 mg, 0.38 mmol), N-methylglycine ethyl ester 3b (26.8 mg, 0.23 mmol), and CuI (1.8 mg, 0.0095 mmol) in dioxane (0.5 mL) was stirred at 170 °C for 1 h under the microwave irradiation (300 W). After the three-component indole formation was completed monitored by TLC, MsOH (1 mL) was added at rt and the mixture was stirred at 80 °C for 30 min. The reaction mixture was diluted with H$_2$O followed by neutralization with saturated aqueous NaHCO$_3$. The aqueous solution was extracted with EtOAc (twice). The organic layer was washed with brine and dried over MgSO$_4$. The filtrate was concentrated under reduced pressure to leave an oily residue, which was purified by column chromatography over silica gel with CHCl$_3$/CH$_3$OH (50:1) as the eluent to give 7a (32.6 mg, 72%) as a pale yellow solid, which was recrystallized from CHCl$_3$–hexane: colorless crystals: mp 183 °C; IR (neat) 1,639 cm^{-1} (C=O); ^{1}H NMR (500 MHz, CDCl$_3$) δ 2.55 (s, 3H, 2-NMe), 3.26 (s, 2H, 3-CH$_2$), 3.63 (s, 3H, 9-NMe), 3.76 (s, 2H, 1-CH$_2$), 7.27–7.29 (m, 3H, Ar), 8.18–8.20 (m, 1H, Ar); ^{13}C NMR (125 MHz, CDCl$_3$) δ 30.0, 45.1, 50.4, 63.3, 109.3, 110.7, 121.5, 122.7, 123.2, 124.1, 137.6, 149.8, 189.9; MS (FAB) m/z (%): 215 (MH$^+$, 100); HRMS (FAB) calcd for C$_{13}$H$_{15}$N$_2$O (MH$^+$): 215.1184; found: 215.1180.

3.1.11 2-Allyl-2,3-dihydro-1H-pyrido[3,4-b]indol-4(9H)-one (7b)

By a procedure similar to that described for 7a, 1i (25.0 mg, 0.19 mmol) was converted into 7b (35.1 mg, 77%) by treatment with N-allylglycine ethyl ester 3c (26.8 μL, 0.23 mmol); colorless crystals (from CHCl$_3$–hexane): mp 116 °C; ^{1}H NMR (500 MHz, CDCl$_3$) δ 3.29 (d, 2H, $J = 6.9$ Hz, NCH$_2$CH), 3.32 (s, 2H, COCH$_2$), 3.63 (s, 3H, Me), 3.82 (s, 2H, ArCH$_2$), 5.30–5.24 (m, 2H, CH=CH$_2$), 5.86–5.94 (m, 1H, CH), 7.26–7.28 (m, 3H, Ar), 8.12–8.20 (m, 1H, Ar); ^{13}C NMR (125 MHz, CDCl$_3$) δ 30.0, 47.9, 60.2, 61.3, 109.3, 111.2, 119.0, 121.5, 122.7, 123.2, 124.1, 134.1, 137.6, 149.7, 189.9; MS (FAB) m/z (%): 241 (MH$^+$, 30); HRMS (FAB) calcd for C$_{15}$H$_{17}$N$_2$O (MH$^+$): 241.1341; found: 241.1336.

3.1.12 2-Butyl-2,3-dihydro-1H-pyrido[3,4-b] indol-4(9H)-one (7c)

By a procedure similar to that described for **7a**, **1i** (25.0 mg, 0.19 mmol) was converted into **7c** (34.2 mg, 68%) by treatment with N-butylglycine ethyl ester **3d** (33.4 mg, 0.21 mmol); colorless crystals (from CHCl$_3$–hexane): mp: 109 °C: IR: $\tilde{v}$ = 1,650 cm^{-1} (CO); ^{1}H NMR (500 MHz, CDCl$_3$) δ 0.94 (t, 3H, J = 7.4 Hz, CH$_2$CH$_3$), 1.34–1.41 (m, 2H, CH$_2$CH$_3$) 1.54–1.60 (m, 2H, NCH$_2$CH$_2$), 2.64 (t, 2H, J = 7.4 Hz, NCH$_2$CH$_2$), 3.30 (s, 2H, COCH$_2$), 3.64 (s, 3H, NMe), 3.82 (s, 2H, ArCH$_2$), 7.26–7.28 (m, 3H, Ar), 8.16–8.18 (m, 1H, Ar); ^{13}C NMR (125 MHz, CDCl$_3$) δ 13.9, 20.4, 29.2, 30.0, 48.7, 57.1, 61.3, 109.3, 111.1, 121.5, 122.7, 123.2, 124.1, 137.6, 150.0, 190.1; MS (FAB) m/z (%): 257 (MH$^+$, 100); HRMS (FAB) calcd for C$_{16}$H$_{21}$N$_2$O (MH$^+$): 257.1654; found: 257.1660.

3.1.13 General Procedure for Synthesis of 1,2,3,4-Tetrahydro-β-carboline by Domino Copper-Catalyzed Three-Component Indole Formation and Cyclization by MsOH: Synthesis of 2-Benzyl-2,3-dihydro1H-pyrido[3,4-b]indol-4(9H)-one (7d) (Conditions B)

A mixture of N-methyl-2-ethynylaniline **1i** (25.0 mg, 0.19 mmol), paraformaldehyde (11.4 mg, 0.38 mmol), N-benzylglycine ethyl ester **3f** (44.2 mg, 0.23 mmol), and CuBr (1.3 mg, 0.0095 mmol) in dioxane (0.5 mL) was stirred for 15 min at 120 °C and additionally for 15 min at 140 °C, using the microwave apparatus. After the three-component indole formation was completed (monitored by TLC, MsOH (1 mL) was added at rt and the mixture was stirred for 30 min at 80 °C. The reaction mixture was diluted with H$_2$O followed by neutralization with saturated aqueous NaHCO$_3$. The aqueous solution was extracted with EtOAc (twice). The organic layer was washed with brine and dried over MgSO$_4$. The filtrate was concentrated under reduced pressure to give an oily residue, which was purified by column chromatography over silica gel with hexane/AcOEt (2:1 to 1:1) as the eluent to give **7d** (31.5 mg, 57%) as a yellow pale solid which was recrystallized from CHCl$_3$-hexane. colorless crystals: mp 156 °C; IR: $\tilde{v}$ = 1,650 cm^{-1} (CO); ^{1}H NMR (500 MHz, CDCl$_3$) δ 3.38 (s, 2H, COCH$_2$), 3.60 (s, 3H, NMe), 3.82 (s, 2H, ArCH$_2$), 3.84 (s, 2H, ArCH$_2$), 7.25–7.35 (m, 8H, Ar), 8.18–8.20 (m, 1H, Ar); ^{13}C NMR (CDCl$_3$) δ 30.1, 48.0, 61.50, 61.53, 109.3, 111.1, 121.6, 122.8, 123.3, 124.1, 127.7, 128.6 (2C), 129.1 (2C), 137.0, 137.6, 149.7, 190.1; MS (FAB) m/z (%): 291 (MH$^+$, 35); HRMS (FAB) calcd for C$_{19}$H$_{19}$N$_2$O (MH$^+$): 291.1497; found: 291.1504.

3.1.14 (R)-2,3-Dimethyl-2,3-dihydro-1H-pyrido[3,4-b]indol-4(9H)-one (7e) (Conditions C)

By a procedure similar to that described for **7d**, **1i** (25.0 mg, 0.19 mmol) was converted into **7e** [27.2 mg, 63, 95% ee (Chiralcel OD-H with a linear gradient of i-PrOH (20–40% over 45 min) in hexane in the presence of 0.1% Et$_2$NH)] by treatment with N-methylalanine methyl ester **3g** (26.8 mg, 0.23 mmol) and by the reaction at 120 °C at the indole formation step; colorless crystals (from CHCl$_3$–hexane): mp 143 °C; $[\alpha]_D^{24}$ 15.5 (c 0.67, CHCl$_3$); IR: $\tilde{\nu}$ = 1,647 cm^{-1} (CO); ^{1}H NMR (CDCl$_3$) δ 1.36 (d, 3H, J = 7.0 Hz, 3-CH$_3$), 2.59 (s, 3H, 2-CH$_3$), 3.31 (q, 1H, J = 7.0 Hz, CH), 3.69 (s, 3H, 9-CH$_3$), 3.87 (d, 1H, J = 16.6 Hz, CHH), 4.14 (d, 1H, J = 16.6 Hz, CHH), 7.27–7.33 (m, 3H, Ar), 8.19–8.22 (m, 1H, Ar); ^{13}C NMR (CDCl$_3$) δ 12.2, 30.0, 42.4, 47.3, 65.0, 109.2, 109.4, 121.6, 122.7, 123.2, 124.6, 137.7, 148.0, 193.6.; MS (FAB) m/z (%): 229 (MH$^+$, 100); HRMS (FAB) calcd for C$_{14}$H$_{17}$N$_2$O (MH$^+$): 229.1341; found: 229.1334.

3.1.15 (R)-3-Isobutyl-2-methyl-2,3-dihydro-1H-pyrido[3,4-b]indol-4(9H)-one (7f)

By a procedure similar to that described for **7e**, **1i** (25.0 mg, 0.19 mmol) was converted into **7f** (19.0 mg, 37%) by treatment with N-methylleucine methyl ester **3h** (36.4 mg, 0.23 mmol); colorless crystals (from CHCl$_3$–hexane): mp 157 °C; $[\alpha]_D^{24}$ −13.2 (c 0.67, CHCl$_3$); IR: $\tilde{\nu}$ = 1,646 cm^{-1} (CO); ^{1}H NMR (500 MHz, CDCl$_3$), δ 0.95 (d, 3H, J = 6.9 Hz, CHCH$_3$CH_3), 0.99 (d, 3H, J = 6.9 Hz, CHCH$_3$CH_3), 1.49–1.54 (m, 1H, CHCHH), 1.60–1.65 (m, 1H, CHCHH), 1.85–1.93 (m, 1H, CHCH$_3$CH$_3$), 2.59 (s, 3H, 2-CH$_3$), 3.26 (dd, 1H, J = 8.6, 6.3 Hz, CH), 3.70 (s, 3H, 9-CH$_3$), 3.90 (d, 2H, J = 17.2, 1-CHH), 4.35 (d, 2H, J = 17.2, 1-CHH), 7.27–7.35 (m, 3H, Ar), 8.19–8.22 (m, 1H, Ar); ^{13}C NMR (125 MHz, CDCl$_3$) δ 22.1, 23.0, 25.2, 30.1, 37.4, 43.0, 45.8, 67.7, 109.1, 109.3, 121.7, 122.7, 123.0, 124.6, 137.6, 146.5, 195.2; MS (FAB) m/z (%): 271 (MH$^+$, 65); HRMS (FAB) calcd for C$_{17}$H$_{23}$N$_2$O (MH$^+$): 271.1810; found: 271.1804.

3.1.16 (R)-3-Benzyl-2-methyl-2,3-dihydro-1H-pyrido[3,4-b]indol-4(9H)-one (7g)

By a procedure similar to that described for **7e**, **1i** (25.0 mg, 0.19 mmol) was converted into **7g** (26.8 mg, 46%) as by treatment with N-methylphenylalanine methyl ester **3i** (44.2 mg, 0.23 mmol); mp 94 °C by (CHCl$_3$–hexane); $[\alpha]_D^{24}$ −62.8 (c 0.67, CHCl$_3$); IR: $\tilde{\nu}$ = 1,646 cm^{-1} (CO); ^{1}H NMR (500 MHz, CDCl$_3$) δ 2.54

(s, 3H, 2-CH$_3$), 3.00 (dd, 1H, J = 14.3, 9.2 Hz, CHCHH), 3.12 (dd, 1H, J = 14.3, 5.2 Hz, CHCHH), 3.54 (dd, 1H, J = 9.2, 5.2 Hz, CH), 3.69 (s, 3H, 9-CH$_3$), 3.91 (d, 1H, J = 17.2 Hz, 1-CHH), 4.37 (d, 1H, J = 17.2 Hz, 1-CHH), 7.18–7.36 (m, 8H, Ar), 8.22–8.25 (m, 1H, Ar); ^{13}C NMR (125 MHz, CDCl$_3$) δ 30.1, 34.7, 43.0, 46.5, 71.0, 109.3, 109.7, 121.7, 122.8, 123.2, 124.5, 126.1, 128.3 (2C), 129.0 (2C), 137.7, 139.3, 147.1, 193.5; MS (FAB) m/z (%): 305 (MH$^+$, 45); HRMS (FAB) calcd for C$_{20}$H$_{21}$N$_2$O (MH$^+$): 305.1654; found: 305.1649.

3.1.17 5,6,8,9,10,11,11a,12-Octahydroindolo[3,2-b]quinolizine (7h)

By a procedure similar to that described for **7e**, **1i** (25.0 mg, 0.19 mmol) was converted into **7h** (14.0 mg, 29%) as an yellow oil by treatment with methyl pipecolinate **3j** (44.2 mg, 0.23 mmol); IR: $\tilde{\nu}$ = 1,646 cm^{-1} (CO); ^{1}H NMR (500 MHz, CDCl$_3$) δ 1.38–1.78 (m, 4H, 4 × CHH), 1.92 (m, 1H, CHH), 2.51 (m, 2H, 2 × CHH), 2.83 (m, 1H, CHH), 3.10–3.13 (m, 1H, CH), 3.67–3.70 (m, 4H, 9-Me, ArCHH), 4.08 (d, 1H, J = 16.0 Hz, ArCHH), 7.27–7.32 (m, 3H, Ar), 8.18–8.22 (m, 1H, Ar); ^{13}C NMR (125 MHz, CDCl$_3$) δ 24.0, 25.4, 26.5, 29.9, 50.5, 56.3, 67.5, 109.2, 110.4, 121.6, 122.7, 123.2, 124.5, 137.6, 149.0, 191.0; MS (FAB) m/z (%): 255 (MH$^+$, 100); HRMS (FAB) calcd for C$_{16}$H$_{19}$N$_2$O (MH$^+$): 255.1497; found: 255.1507.

References

1. Pless G, Frederiksen TJ, Garcia JJ, Reiter RJ (1999) J Pineal Res 26:236–246
2. Herraiz T, Galisteo J (2002) Free Radic Res 36:923–928
3. Ichikawa M, Ryu K, Yoshica J, Ide N, Yoshida S, Sasaoka T, Sumi S (2002) Biofactors 16:57–72
4. Herraiz T, Galisteo J, Chamorro CJ (2003) J Agric Food Chem 51:2168–2173
5. Herraiz T, Galisteo J (2003) J Agric Food Chem 51:7156–7161
6. Bi W, Bai L, Cai J, Liu S, Peng S, Fischer NO, Tok JB-H, Wang G (2006) Bioorg Med Chem Lett 16:4523–4527
7. Bi W, Cai J, Liu S, Baudy-Floc'h M, Bi L (2007) Bioorg Med Chem 15:6906–6919
8. Yu P, Wang T, Li J, Cook JM (2000) J Org Chem 65:3173–3191
9. Zhou H, Liao X, Cook JM (2004) Org Lett 6:249–252
10. Liu C, Masuno MN, MacMillan JB, Molinski TF (2004) Angew Chem Int Ed 43:5951–5954
11. Yamashita T, Kawai N, Tokuyama H, Fukuyama T (2005) J Am Chem Soc 127:15038–15039
12. Yu J, Wearing XZ, Cook JM (2005) J Org Chem 70:3963–3979
13. Zhou H, Han D, Liao X, Cook JM (2005) Tetrahedron Lett 46:4219–4224
14. Zhou H, Liao X, Yin W, Ma J, Cook JM (2006) J Org Chem 71:251–259
15. Ma J, Yin W, Zhou H, Cook JM (2007) Org Lett 9:3491–3494
16. Volz F, Krause N (2007) Org Biomol Chem 5:1519–1521
17. Mergott DJ, Zuend SJ, Jacobsen EN (2008) Org Lett 10:745–748

18. Martin SF, Chen KX, Eary CT (1999) Org Lett 1:79–81
19. Neipp CE, Martin SF (2003) J Org Chem 68:8867–8878
20. Ohba M, Natsutani I, Sakuma T (2004) Tetrahedron Lett 45:6471–6474
21. Ohba M, Natsutani I, Sakuma T (2007) Tetrahedron 63:10337–10344
22. Czarnocki SJ, Wojtasiewicz K, Jóźwiak AP, Maurin JK, Czarnocki Z, Drabowicz J (2008) Tetrahedron 64:3176–3182
23. Shankaraiah N, da Silva WA, Andrade CKZ, Santos LS (2008) Tetrahedron Lett 49:4289–4291
24. Abramovitch R A, Shapiro D (1956) J Chem Soc 4529–4589
25. Pelchobicz Z, Bergmann ED (1959) J Chem Soc 847
26. Frangatos G, Kohan G, Chubb FL (1960) Can J Chem 38:1082–1086
27. Wender PA, White AW (1983) Tetrahedron 39:3767–3776
28. Luis SV, Burguete MI (1991) Tetrahedron 47:1737–1744
29. Dantale SW, Söderberg BCG (2003) Tetrahedron 59:5507–5514
30. Baruah B, Dasu K, Vaitilingam B, Mamnoor P, Venkata PP, Rajagopal S, Yeleswarapu KR (2004) Bioorg Med Chem 12:1991–1994
31. Iwadate M, Yamashita T, Tokuyama H, Fukuyama T (2005) Heterocycles 66:241–249
32. Ohno H, Ohta Y, Oishi S, Fujii N (2007) Angew Chem Int Ed 46:2295–2298
33. Ohta Y, Chiba H, Oishi S, Fujii N, Ohno H (2009) J Org Chem 74:7052–7058
34. Ohta Y, Oishi S, Fujii N, Ohno H (2008) Chem Commun 835–837
35. Ohta Y, Chiba H, Oishi S, Fujii N, Ohno H (2008) Org Lett 10:3535–3838
36. Rubiralta M, Diez A, Bosch J, Solans X (1989) J Org Chem 54:5591–5597
37. Murakami Y, Yokoyama Y, Aoki C, Miyagi C, Watanabe T, Ohmoto T (1987) Heterocycles 26:875–878
38. Suzuki H, Yokoyama Y, Miyagi C, Murakami Y (1991) Chem Pharm Bull 39:2170–2172
39. Murakami Y, Yokoyama Y, Aoki C, Suzuki H, Sakurai K, Shinohara T, Miyagi C, Kimura Y, Takahashi T, Watanabe T, Ohmoto T (1991) Chem Pharm Bull 39:2189–2195
40. Suzuki H, Iwata C, Sakurai K, Tokumoto K, Takahashi H, Hanada M, Yokoyama Y, Murakami Y (1997) Tetrahedron 53:1593–1606
41. Suzuki H, Umemoto M, Hagiwara M, Ohyama T, Yokoyama Y, Murakami Y (1999) J Chem Soc, Perkin Trans 1:1717–1723
42. Jennings LD, Foreman KW, Rush TS III, Tsao DHH, Mosyak L, Li Y, Sukhdeo MN, Ding W, Dushin EG, Kenny CH, Moghazeh SL, Petersen PJ, Ruzin AV, Tuckman M, Sutherland AG (2004) Bioorg Med Chem Lett 14:1427–1431
43. Karpov AS, Oeser T, Müller TJJ (2004) Chem Commun 1502–1503
44. Karpov AS, Rominger F, Müller TJJ (2005) Org Biomol Chem 3:4382–4391
45. Kabalka GW, Wang L, Pagni RM (2001) Tetrahedron 57:8017–8028
46. Gribble GW, Saulnier MG (1983) J Org Chem 48:607–609
47. Kurisaki T, Naniwa T, Yamamoto H, Imagawa H, Nishizawa M (2007) Tetrahedron Lett 48:1871–1874
48. Yoo EJ, Chang S (2008) Org Lett 10:1163–1166
49. Robichaud J, Tremblay F (2006) Org Lett 8:597–600
50. Webert J-W, Cagniant D, Cagniant P, Kirsch G, Weber J-V (1983) J Hetercycl Chem 20:49–53
51. Reichwein JF, Liska RMJ (2000) Eur J Org Chem 2335–2344
52. Zuliani V, Carmi C, Rivara M, Fantini M, Lodola A, Vacondio F, Bordi F, Plazzi PV, Cavazzoni A, Galetti M, Alfieri RR, Petronini PG, Mor M (2009) Eur J Med Chem 44:3471–3479
53. Hu C, Chen Z, Yang G (2004) Synth Commun 34:219–224
54. Fang JB, Sanghi R, Kohn J, Goldman AS (2004) Inorg Chim Acta 357:2415–2426
55. Wen S-J, Hu T-S, Yao Z-J (2005) Tetrahedron 61:4931–4938
56. Adima A, Bied C, Moreau JJE, Man MWC (2004) Eur J Org Chem 2582–2588
57. Tong STA, Barker D (2004) Tetrahedron Lett 47:5017–5020

Chapter 4
Concise Synthesis of Indole-Fused 1,4-Diazepines through Copper(I)-Catalyzed Domino Three-Component Coupling-Cyclization-N-Arylation under Microwave Irradiation

Tandem catalysis [1–10], which involves several catalytic cycles within the same medium to produce a desired product, is becoming increasingly important for the economic and environmental acceptability of the process. Copper salts are efficient catalysts in various transformations, including formation of carbon–carbon and carbon–nitrogen bonds [11–14]. The author postulated they could play key parts in construction of complex nitrogen heterocycles with important biological activities through formation of multiple bonds [15–23].

Indole and 1,4-benzodiazepine frameworks are useful templates for drug discovery. Indole-fused 1,4-diazepine [24–31], found in various bioactive compounds, can also be an attractive drug template. In Scheme 1, the author reported a novel

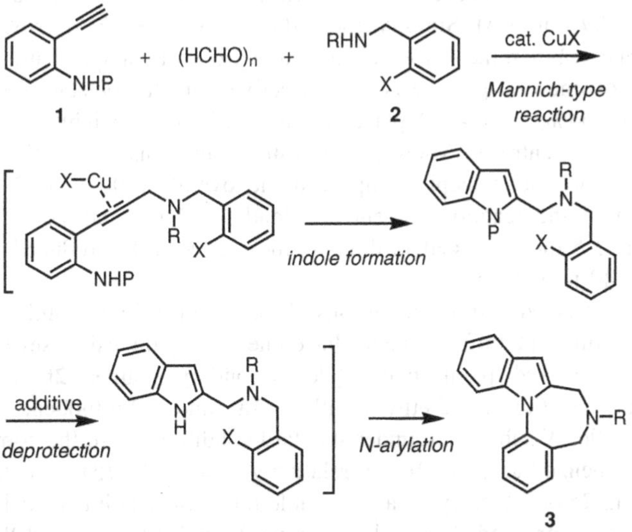

Scheme 1 Copper(I)-catalyzed domino three-component coupling-cyclization-N-arylation reaction

Y. Ohta, *Copper-Catalyzed Multi-Component Reactions*, Springer Theses,
DOI: 10.1007/978-3-642-15473-7_4, © Springer-Verlag Berlin Heidelberg 2011

copper(I)-catalyzed synthesis of 2-(aminomethyl)indoles via a three-component coupling-cyclization reaction [32, 33]. This new indole-forming reaction prompted the author to develop a novel method for the synthesis of indole-fused tetracyclic compounds by three-component indole formation and simultaneous copper-catalyzed N-arylation (Scheme 1). The author expected that a copper salt could catalyze multiple transformations, including Mannich-type coupling of ethynylaniline derivative 1 with formaldehyde and N-substituted o-halobenzylamine 2, indole formation, and arylation of the indole nitrogen. In this section, the author reports a direct access to indole-fused tetracyclic compounds 3 containing the 1,4-diazepine framework by copper(I)-catalyzed domino reactions, which involve the formation of one carbon–carbon bond and three carbon–nitrogen bonds.

The author chose N-mesyl-2-ethynylaniline 1a as a model substrate because three-component indole formation requires N-substituted ethynylanilines [32]. Appropriate conditions were initially investigated for one-pot three-component indole formation, deprotection of the mesyl group, and subsequent N-arylation. A mixture of 1a, paraformaldehyde (2 equiv), and secondary amine 2a (1.1 equiv) was treated with CuI (5 mol%) in toluene and, after indole formation was completed (monitored by TLC), an additive for cleavage of the N-mesyl group was introduced (Table 1) (One portion addition of all the reactants including the alkoxide at the beginning of the reaction caused decomposition of the starting material). Addition of MeOK and heating of the reaction mixture under reflux for 1 h promoted the desired arylation of the indole nitrogen to afford the expected tetracyclic compound 3a [34] in ca. 43% yield (entry 1). t-BuOK was less effective, leading to ca. 38% yield of 3a (entry 2). These runs furnished tetracyclic compound 3a containing some impurities that were not easily removed, but the reaction with MeONa under reflux for 3 h gave pure 3a in 51% yield after column chromatography (entry 3). Simultaneous addition of racemic trans-N,N'-dimethylcyclohexane-1,2-diamine, an efficient ligand for CuI-catalyzed intermolecular N-arylation of indoles [35], was not effective for the present formation of 1,4-diazepine (34%, entry 4). Replacement of CuI by CuBr slightly decreased the yield of 3a (49%, entry 5). Microwave-assisted conditions at 170 °C for the formation of indole and diazepine improved the overall yield to 64% (entry 6). Investigation of the reaction solvent and loading of the catalyst (entries 6–9) revealed that 2.5 mol% of CuI in dioxane most effectively produced 3a in 88% yield within 40 min (entry 9).

Having established optimal conditions (Table 1, entry 9), the author examined the scope of this indole-fused benzodiazepine formation using several 2-ethynylanilines 1a–e and paraformaldehyde, secondary amines 2b–d (Table 2). Whereas the reaction of 2-ethynylaniline 1a, and 2-bromobenzylamine 2b bearing a smaller N-substituent under standard conditions gave the corresponding indole-fused benzodiazepine 3b in relatively low yield (51%, entry 1), the reaction using 2c or 2d, carrying a removable nitrogen substituent such as benzyl and allyl groups, proceeded smoothly to give 3c and 3d in 83 and 81% yields, respectively (entries 2 and 3). Ethynylaniline 1b bearing a methoxycarbonyl group at the para-position of the amino group gave a poor result to afford 3e

Table 1 Screening of reaction conditions using ethynylaniline **1a** and secondary amine **2a**

Entry	Catalyst (mol%)	Solvent	Conditions A[a]	Additive (equiv)	Conditions B[a]	Yield[b] (%)
1	CuI (5)	Toluene	Reflux, 6 h	MeOK (6)	Reflux, 1 h	43
2	CuI (5)	Toluene	Reflux, 6 h	t-BuOK (6)	Reflux, 0.5 h	38
3	CuI (5)	Toluene	Reflux, 6 h	MeONa (6)	Reflux, 3 h	51
4	CuI (5)	Toluene	Reflux, 6 h	MeONa (6) ligand (0.1)[c]	80 °C, 4 h	34
5	CuBr (5)	Toluene	Reflux, 6 h	MeONa (6)	Reflux, 3 h	49
6	CuI (5)	Toluene	MW, 170 °C, 20 min	MeONa (6)	MW, 170 °C, 20 min	64
7	CuI (5)	Dioxane	MW, 170 °C, 20 min	MeONa (6)	MW, 170 °C, 20 min	81
8	CuI (1)	Dioxane	MW, 170 °C, 20 min	MeONa (6)	MW, 170 °C, 20 min	77
9	CuI (2.5)	Dioxane	MW, 170 °C, 20 min	MeONa (6)	MW, 170 °C, 20 min	88

After the reactions with 2-ethynylaniline **1a**, paraformaldehyde (2 equiv), and secondary amine **2a** (1.1 equiv) was completed on TLC, additives were introduced
[a] *MW* microwave irradiation
[b] Isolated yields
[c] Ligand = (±)-*trans*-N,N'-dimethylcyclohexane-1,2-diamine

(23% yield), along with a complex mixture of unidentified products (entry 4) (since the formation 2-(aminomethyl)indole using **1b** and **2d** proceeded efficiently (quantitative yield), deprotection conditions using MeONa caused undesired side reactions). Anilines **1c** and **1d** with a *para*-trifluoromethyl or methyl group, respectively, were good substrates for this copper-catalyzed reaction sequence (entries 5 and 6). The reaction with ethynylaniline **1e** containing a trifluoromethyl group at the *meta*-position gave a moderate yield of **3 h** (53% yield, entry 7). Thus, the copper-catalyzed synthesis of indole-fused benzodiazepine was applicable to various N-substituted o-bromobenzylamines and 2-ethynylanilines with an electron-donating or electron-withdrawing group.

Synthesis of tetracyclic compounds containing a heterocycle-fused 1,4-diazepine was investigated (Scheme 2). By employing the secondary amines **4** and **6** involving a pyridine and thiophene moiety, respectively, the reaction directly delivered the desired pyridine- and thiophene-fused tetracyclic compounds **5** and **7**

Table 2 Construction of tetracyclic compounds using substituted ethynylanilines and o-bromobenzylamines

Entry	Ethynylaniline	Secondary amine	Product (%)[b]
1	1a	MeHN— 2b	3b (51)
2	1a	BnHN— 2c	3c (83)
3	1a	allyl—N—H 2d	3d (81)
4	1b: R = CO$_2$Me	2d	3e (R = CO$_2$Me, 23)
5	1c: R = CF$_3$	2d	3f (R = CF$_3$, 81)
6	1d: R = Me	2d	3g (R = Me, 85)
7	1e	allyl—N—H 2d	3h (53)

All reactions were conducted with ethynylaniline **1**, paraformaldehyde (2 equiv), and secondary amine **2** (1.1 equiv) in the presence of CuI (2.5 mol%) in 1,4-dioxane at 170 °C for 20–40 min under microwave irradiation. After the indole formation was completed (monitored by TLC), MeONa (6 equiv) was added and the mixture was heated at 170 °C for 20 min under microwave irradiation[a] Isolated yields

in 71 and 56% yields, respectively. From these observations, this copper-catalyzed formation of tetracyclic compounds allows the synthesis of indole-fused 1,4-diazepines containing another heterocyclic ring system.

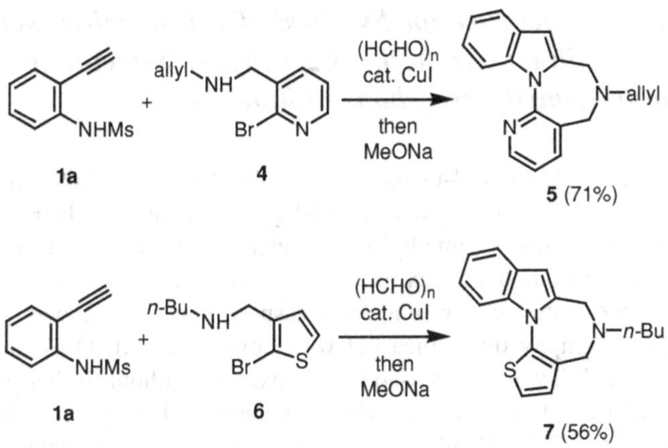

Scheme 2 Direct synthesis of pyridine- or thiophene-fused tetracyclic compounds

In conclusion, the author developed a novel method for the preparation of fused indoles by copper-catalyzed domino three-component coupling-indole formation-N-arylation. Starting from simple 2-ethynylanilines and o-bromobenzylamines, complex indole-fused tetracyclic compounds were easily and directly synthesized in a single reaction vessel. This is the first example of copper-catalyzed one-pot reaction including three catalytic cycles and formation of four bonds.

4.1 Experimental Section

The compounds **1a** [36], **2a, c, d** [37] are known.

The compound **2b**, 2-bromo-3-(bromomethyl)thiophene, and 2-bromopicolinaldehyde are commercially available.

4.1.1 General Methods

Exact mass (HRMS) spectra were recorded on JMS-HX/HX 110A mass spectrometer. ^{1}H NMR spectra were recorded using a JEOL AL-500 spectrometer at 500 MHz frequency. Chemical shifts are reported in δ (ppm) relative to Me$_4$Si (in CDCl$_3$,) as internal standard. ^{13}C NMR spectra were recorded using a JEOL AL-500 and referenced to the residual CHCl$_3$ signal. Microwave reaction was conducted in a sealed glass vessel (capacity 10 mL) using CEM Discover microwave reactor with a run time of no more than 10 min. The temperature was monitored using IR sensor mounted under the reaction vessel. For column chromatography, Wakosil C-300 was employed.

4.1.2 General Procedure for Synthesi of 2-Ethynyl-N-methane-sufonylaniline: Synthesis of 2-Ethynyl-N-methane sulfonyl-4-methoxycarbonylaniline (1b)

To the mixture of 2-bromo-4-methoxycarbonylaniline (2 g, 8.69 mmol), PdCl$_2$ (PPh$_3$)$_2$ (0.15 g, 0.22 mmol), and CuI (0.04 g, 0.22 mmol) in THF (2 mL) and Et$_3$N (20 mL) was added trimethylsilylacetylene (1.42 mL, 10.43 mmol) at rt under Ar. After stirred under reflux for 16 h, the reaction mixture was filtered over Celite and concentrated under reduced pressure. The residue was purified by column chromatography over silica gel with hex/AcOEt (10:1) as the eluent to give a colorless solid, which was used in the next step without further purification. To a stirred solution of this TMS-acetylenated compound in pyridine (20 mL) was added dropwise Ms-Cl (0.44 mL, 6.79 mmol) at 0 °C under Ar. After stirred at rt for 12 h, the reaction mixture was quenched with aqueous saturated NaHCO$_3$ and extracted with EtOAc. The organic layer was washed with 1 N HCl, aqueous saturated NaHCO$_3$, and brine, dried over MgSO$_4$, and concentrated under reduced pressure. The residue was purified by column chromatography over silica gel with hex/AcOEt (1:1) to give a colorless solid, which was used in the next step without further purification. To the stirred solution of this mesylate in THF (7 mL) was added dropwise TBAF (2.2 mL, 1 M in THF, 2.2 mmol) at 0 °C. After stirred for 5 min at this temperature, the reaction mixture was quenched with aqueous sat-urated citric acid and extracted with EtOAc. The organic layer was washed with H$_2$O, aqueous saturated NaHCO$_3$, and brine, drid over MgSO$_4$, and concentrated under reduced pressure. The residue was purified by column chromatography over silica gel with hex/AcOEt (3:1 to 1:1) to give a colorless solid, which was recrystallized from hex–AcOEt to give pure 1b (0.49 g, 22% over 3 steps) as colorless crystals: m.p. 122 °C; IR (neat) 2106 cm^{-1} (C≡C); ^{1}H NMR (500 MHz, CDCl$_3$) δ 3.10 (s, 3H, SO$_2$CH$_3$), 3.55 (s, 1H,C≡CH), 3.92 (s, 3H, OMe), 7.29 (br, 1H, NH), 7.67 (d, $J = 8.8$ Hz, 1H, Ar), 8.04 (dd, $J = 8.8$, 2.0 Hz, 1H, Ar), 8.18 (d, $J = 2.0$ Hz, Ar); ^{13}C NMR (125 MHz, CDCl$_3$) δ 40.3, 52.3, 77.6, 85.8, 111.5, 116.8, 125.8, 131.8, 134.5, 142.3, 165.5. Anal. Calcd for C$_{11}$H$_{11}$NO$_4$S: C, 52.16; H, 4.38; N, 5.53. Found: C, 52.11; H, 4.22; N, 5.50.

4.1.3 2-Ethynyl-N-methanesulfonyl-4-trifluoromethyl carbonylaniline (1c)

By a procedure similar to that described for 1b, 2-iodo-4-trifluoromethylaniline (1.50 g, 5.23 mmol) was converted into 1c (0.47 g, 34% over 3 steps); colorless crystals (from AcOEt–hexane): m.p. 92 °C; IR (neat) 2111 cm^{-1} (C≡C); ^{1}H NMR (500 MHz, CDCl$_3$) δ 3.11 (s, 3H, SO$_2$CH$_3$), 3.61 (s, 1H, C≡CH), 7.30 (br, 1H, NH), 7.63 (dd, $J = 8.7$, 1.7 Hz, 1H, Ar), 7.74 (d, $J = 8.7$ Hz, 1H, Ar), 7.76 (d, $J = 1.7$ Hz, 1H, Ar); ^{13}C NMR (125 MHz, CDCl$_3$) δ 40.3, 77.3, 86.4, 112.2,

117.8, 123.3 (q, $J = 272.3$ Hz), 126.4 (q, $J = 33.6$ Hz), 127.4 (q, $J = 3.6$ Hz), 130.0 (q, $J = 3.6$ Hz), 141.5. *Anal.* Calcd. for $C_{10}H_8F_3NO_2S$: C, 45.63; H, 3.06; N, 5.32. Found C, 45.67; H, 3.07; N, 5.29.

4.1.4 2-Ethynyl-N-methanesulfonyl-4-methylaniline (1d)

By a procedure similar to that described for **1b**, 2-iodo-4-methylaniline (2.03 g, 5.23 mmol) was converted into **1c** (1.53 g, 84% over 3 steps); colorless crystals (from AcOEt–hexane): m.p. 95 °C; IR (neat) 2100 cm^{-1} (C≡C); ^{1}H NMR (500 MHz, CDCl$_3$) δ 2.31 (s, 3H, ArCH$_3$), 2.98 (s, 3H, SO$_2$CH$_3$), 3.45 (s, 1H, C≡CH), 6.88 (br, 1H, NH), 7.19 (dd, $J = 8.4, 1.9$ Hz, 1H, Ar), 7.32 (d, $J = 1.9$ Hz, 1H, Ar), 7.49 (d, $J = 8.4$ Hz, 1H, Ar); ^{13}C NMR (125 MHz, CDCl$_3$) δ 20.5, 39.4, 79.0, 84.2, 113.3, 120.5, 131.3, 133.1, 134.9, 135.9. *Anal.* Calcd. for $C_{10}H_{11}NO_2S$: C,57.39; H, 5.19; N, 6.69. Found C, 57.39; H, 5.30; N, 6.69.

4.1.5 2-Ethynyl-N-methanesulfonyl-5-trifluoromethyl carbonylaniline (1e)

By a procedure similar to that described for **1b**, 2-bromo-5-trifluoroaniline (2.09 g, 8.69 mmol) was converted into **1c** (0.35 g, 15% over 3 steps); colorless crystals (from AcOEt–hexane): m.p. 107 °C; IR (neat) 2113 cm^{-1} (C≡C); 3.08 (s, 3H, SO$_2$CH$_3$), 3.63 (s, 1H, C≡CH), 7.17 (br, 1H, NH), 7.38 (dd, $J = 8.0, 0.6$ Hz, 1H, Ar), 7.62 (d, $J = 8.0$ Hz, 1H, Ar), 7.88 (d, $J = 0.6$ Hz, 1H, Ar); ^{13}C NMR (125 MHz, CDCl$_3$) δ 40.2, 77.5, 87.0, 115.6 (q, $J = 3.6$ Hz), 115.8, 121.0 (q, $J = 3.6$ Hz), 123.2 (q, $J = 272.3$ Hz), 132.5 (q, $J = 33.6$ Hz), 133.4, 139.1. *Anal.* Calcd. for $C_{10}H_8F_3NO_2S$: C, 45.63; H, 3.06; N, 5.32. Found C, 45.68; H, 3.04; N, 5.36.

4.1.6 General Procedure for Synthesis of Indole-Fused 1,4-Diazepine through Three-Component Indole Formation-N-Arylation: Synthesis of 7-n-Butyl-7,8-dihydro-6H-benzo[f]indolo[1,2-a][1,4]diazepine (3a)

A mixture of 2-ethynylaniline **1a** (25 mg, 0.13 mmol), paraformaldehyde (7.7 mg, 0.26 mmol), secondary amine **2a** (35 mg, 0.14 mmol), and CuI (0.61 mg, 0.0032 mmol) in dioxane (1 mL) was stirred for 20 min at 170 °C under the microwave irradiation (200 W). After the three-component coupling-cyclization reaction was completed (monitored by TLC), NaOMe (41.4 mg, 0.77 mmol) was added at rt and the mixture was stirred for 20 min at 170 °C under microwave irradiation (200 W). The reaction mixture was concentrated under reduced pressure and purified by column chromatography over silica gel with hexane/EtOAc (3:1) as the eluent to give **3a** (32.8 mg, 88%) as a pale yellow oil; ^{1}H NMR (500 MHz, CDCl$_3$) δ 0.96 (t, J = 7.3 Hz, 3H, CH$_3$), 1.36–1.43 (m, 2H, CH_2CH$_3$), 1.52–1.65 (br, 2H, NCH$_2$CH_2), 2.39–2.70 (br, 2H, NCH$_2$), 3.20–4.05 (br, 4H, 2 × Ar–CH$_2$), 6.56 (s, 1H, 3-H), 7.16–7.23 (m, 2H, Ar), 7.30 (t, J = 7.4 Hz, 1H, Ar), 7.43 (dd, J = 7.4, 1.3 Hz, 1H, Ar), 7.48 (ddd, J = 7.4, 7.4, 1.3 Hz, 1H, Ar), 7.61–7.69 (m, 3H, Ar); ^{13}C NMR (125 MHz, CDCl$_3$) δ 14.1, 20.7, 30.2, 48.5, 54.9, 55.9, 102.2, 110.3, 120.6, 120.8, 122.2, 122.8, 126.0, 128.67, 128.70, 130.6, 131.2, 135.8, 136.1, 138.6; MS (FAB) m/z (%): 291 (MH$^+$, 100); HRMS (FAB) calcd for C$_{20}$H$_{23}$N$_2$ (MH$^+$): 291.1861; found: 291.1869.

4.1.7 7-Methyl-7,8-dihydro-6H-benzo[f]indolo[1,2-a][1,4] diazepine (3b)

By a procedure similar to that described for indole **3a**, **1a** (25.0 mg, 0.13 mmol) was converted into **3b** (16.3 mg, 51%) as an yellow oil by treatment with **2b**; ^{1}H NMR (500 MHz, CDCl$_3$) δ 2.46 (s, 3H, Me), 3.35–3.43 (br, 1H, CHH), 3.48–3.58 (br, 2H, 2 × CHH), 3.71–3.81 (br, 1H, CHH), 6.58 (s, 1H, 3-H), 7.17–7.20 (m, 1H, Ar), 7.21–7.24 (m, 1H, Ar), 7.31–7.34 (m, 1H, Ar), 7.45 (dd, J = 7.4, 1.3 Hz, 1H, Ar), 7.49–7.52 (m, 1H, Ar), 7.62 (d, J = 8.0 Hz, 1H, Ar), 7.67 (d, J = 8.0 Hz, 1H, Ar), 7.69 (d, J = 8.0 Hz, 1H, Ar); ^{13}C NMR (125 MHz, CDCl$_3$) δ 43.8, 50.4, 56.7, 102.3, 110.3, 120.7, 120.9, 122.3, 123.0, 126.2, 128.7, 128.8, 130.4, 131.2, 135.7, 135.8, 138.6; MS (FAB) m/z (%): 249 (MH$^+$, 100); HRMS (FAB) calcd for C$_{17}$H$_{17}$N$_2$ (MH$^+$): 249.1392; found: 249.1400.

4.1.8 7-Benzyl-7,8-dihydro-6H-benzo[f]indolo[1,2-a][1,4]diazepine (3c)

By a procedure similar to that described for indole **3a**, **1a** (25.0 mg, 0.13 mmol) was converted into **3c** (34.4 mg, 83%) as an yellow oil by treatment with **2c**; ^{1}H NMR (500 MHz, CDCl$_3$) δ 3.34–3.63 (br, 3H, 3 × CHH), 3.69 (s, 2H, ArCH$_2$), 3.79–3.87 (br, 1H, CHH), 6.57 (s, 1H, 3-H), 7.16–7.20 (m, 1H, Ar), 7.21–7.24 (m, 1H, Ar), 7.29–7.33 (m, 1H, Ar), 7.35–7.38 (m, 1H, Ar), 7.41–7.45 (m, 1H, Ar), 7.49 (ddd, J = 7.7, 7.7, 1.6 Hz, 1H, Ar), 7.63 (dd, J = 8.2, 0.9 Hz, 1H, Ar), 7.66–7.70 (m, 1H, Ar),; ^{13}C NMR (125 MHz, CDCl$_3$) δ 47.9, 54.6, 60.2, 102.3, 110.3, 120.6, 120.9, 122.2, 122.9, 126.1, 127.3, 128.5 (2C), 128.7, 128.8, 129.3 (2C), 130.6, 131.3, 135.81, 135.84, 138.65, 138.70; MS (FAB) m/z (%): 325 (MH$^+$, 67); HRMS (FAB) calcd for C$_{23}$H$_{21}$N$_2$ (MH$^+$): 325.1705; found: 325.1706.

4.1.9 7-Allyl-7,8-dihydro-6H-benzo[f]indolo[1,2-a][1,4]diazepine (3d)

By a procedure similar to that described for indole **3a**, **1a** (25.0 mg, 0.13 mmol) was converted into **3d** (28.5 mg, 81%) as an yellow oil by treatment with **2d**; ^{1}H NMR (500 MHz, CDCl$_3$) δ 3.20 (dd, J = 6.7, 0.9 Hz, 2H, NCH$_2$CH), 3.30–3.48 (br, 2H, 2 × NCHH), 3.62–3.69 (br, 1H, CHH), 3.89–3.96 (br, 1H, CHH), 5.25 (dd, J = 10.2, 0.9 Hz, 1H, CH=CHH), 5.31 (d, J = 16.6, 1H, CH=CHH), 5.94–5.62 (m, 1H, CH=CH$_2$), 6.56 (s, 1H, 3-H), 7.16–7.20 (m, 1H, Ar), 7.21–7.24 (m, 1H, Ar), 7.32 (dd, J = 7.5, 7.5 Hz, 1H, Ar), 7.44 (d, J = 7.5 Hz, 1H, Ar), 7.50 (dd, J = 7.5, 7.5 Hz, 1H Ar), 7.63 (d, J = 8.0 Hz, 1H, Ar), 7.67 (dd, J = 7.5, 0.7 Hz, 1H, Ar), 7.69 (d, J = 8.0 Hz, 1H, Ar); ^{13}C NMR (125 MHz, CDCl$_3$) δ 47.8, 54.3, 59.0, 102.3, 110.3, 118.4, 120.7, 120.9, 122.3, 123.0, 126.1, 128.7, 128.8, 130.4, 131.3, 135.7, 135.78, 135.80, 138.7; MS (FAB) m/z (%): 275 (MH$^+$, 100); HRMS (FAB) calcd for C$_{19}$H$_{19}$N$_2$ (MH$^+$): 275.1548; found: 275.1549.

4.1.10 7-Allyl-3-methoxycarbonyl-7,8-dihydro-6H-benzo[f] indolo[1,2-a][1,4]diazepine (3e)

By a procedure similar to that described for indole **3a**, **1b** (32.4 mg, 0.13 mmol) was converted into **3e** (9.9 mg, 23%) as an yellow oil; ^{1}H NMR (500 MHz, CDCl$_3$) δ 3.20 (d, J = 6.7 Hz, 2H, NCH$_2$CH), 3.33 (d, J = 12.2 Hz, 1H, NCHH), 3.42 (d, J = 13.9 Hz, 1H, NCHH), 3.68 (d, J = 12.2 Hz, 1H, NCHH), 3.95 (s, 3H, OMe), 3.95–3.98 (m, 1H, NCHH), 5.26 (d, J = 10.2 Hz, 1H, CH=CHH), 5.31 (dd, J = 17.1, 1.5 Hz, 1H, CH=CHH), 5.93–6.01 (m, 1H, CH), 6.65 (s, 1H, 3-H), 7.37 (dd, J = 7.4, 1.0 Hz, 1H, Ar), 7.45 (dd, J = 7.4, 1.4 Hz, 1H, Ar), 7.53 (dd, J = 7.4, 1.4 Hz, 1H, Ar), 7.61 (d, J = 8.7 Hz, 1H, Ar), 7.67 (d, J = 7.4 Hz, 1H, Ar), 7.93 (dd, J = 8.7, 1.6 Hz, 1H, Ar), 8.42 (d, J = 1.6 Hz, 1H, Ar); ^{13}C NMR (125 MHz, CDCl$_3$) δ 47.7, 51.9, 54.1, 59.0, 103.4, 109.9, 118.6, 122.7, 123.0, 123.7, 123.8, 126.8, 128.3, 129.0, 130.4, 131.4, 135.6, 137.2, 138.1, 138.2, 167.9; MS (FAB) m/z (%): 333 (MH$^+$, 25); HRMS (FAB) calcd for C$_{21}$H$_{21}$N$_2$O$_2$ (MH$^+$): 333.1603; found: 333.1606.

4.1.11 7-Allyl-3-trifluoromethyl-7,8-dihydro-6H-benzo[f] indolo[1,2-a][1,4]di-azepine (3f)

By a procedure similar to that described for indole **3a**, **1c** (33.7 mg, 0.13 mmol) was converted into **3f** (34.7 mg, 81%) as an yellow oil; ^{1}H NMR (500 MHz, CDCl$_3$) δ 3.20 (d, J = 6.7 Hz, 2H, NCH$_2$CH), 3.30 (d, J = 11.9 Hz, 1H, NCHH), 3.44 (d, J = 13.6 Hz, 1H, NCHH), 3.68 (d, J = 11.9 Hz, 1H, NCHH), 3.96

(d, J = 13.6 Hz, 1H, NCHH), 5.27 (dd, J = 10.2, 1.7 Hz, 1H, CH=CHH), 5.31 (dd, J = 17.2, 1.6 Hz, 1H, CH=CHH), 5.93–6.01 (m, 1H, CH), 6.64 (s, 1H, 3-H), 7.37 (dd, J = 7.5, 1.2 Hz, 1H, Ar), 7.44–7.47 (m, 2H, Ar), 7.53 (ddd, J = 7.7, 7.7, 1.6 Hz, 1H, Ar), 7.65–7.68 (m, 2H, Ar), 7.96 (d, J = 7.4 Hz, 1H, Ar), 7.96 (s, 1H, Ar), 8.42 (d, J = 1.6 Hz, 1H, Ar); ^{13}C NMR (125 MHz, CDCl$_3$) δ 47.6, 54.1, 59.0, 102.9, 110.5, 118.6 (q, J = 3.6 Hz), 118.7, 119.1 (q, J = 3.6 Hz), 123.0, 123.1 (q, J = 32.4 Hz), 126.2 (q, J = 272.3 Hz), 126.9, 128.1, 129.0, 130.4, 131.4, 135.5, 137.1, 137.5, 138.0; MS (FAB) m/z (%): 343 (MH$^+$, 25); HRMS (FAB) calcd for C$_{20}$H$_{18}$F$_3$N$_2$ (MH$^+$): 343.1422; found: 343.1424.

4.1.12   7-Allyl-3-methyl-7,8-dihydro-6H-benzo[f]indolo[1,2-a][1,4]diazepine (3g)

By a procedure similar to that described for indole **3a**, **1d** (26.7 mg, 0.13 mmol) was converted into **3g** (31.4 mg, 85%) as an yellow oil; ^{1}H NMR (500 MHz, CDCl$_3$) δ 2.47 (s, 3H, Me), 3.19 (d, J = 6.9 Hz, 2H, NCH$_2$CH), 3.26–3.47 (br, 2H, 2 × NCHH), 3.58–3.68 (br, 1H, NCHH), 3.84–3.96 (br, 1H, NCHH), 5.24 (dd, J = 10.2, 0.6 Hz, 1H, CH=CHH), 5.30 (d, J = 17.2 Hz, 1H, CH=CHH), 5.96–6.01 (m, 1H, CH), 6.48 (s, 1H, 3-H), 7.05 (d, J = 8.0 Hz, 1H, Ar), 7.30 (dd, J = 7.4, 7.4 Hz, 1H, Ar), 7.41–7.52 (m, 4H, Ar), 7.67 (d, J = 8.0 Hz, 1H, Ar); ^{13}C NMR (125 MHz, CDCl$_3$) δ 21.4, 47.8, 54.3, 59.0, 101.9, 110.0, 118.4, 120.6, 122.8, 123.8, 126.0, 128.8, 128.9, 130.0, 130.3, 131.2, 134.1, 135.76, 135.78, 138.8; MS (FAB) m/z (%): 289 (MH$^+$, 100); HRMS (FAB) calcd for C$_{20}$H$_{21}$N$_2$ (MH$^+$): 289.1705; found: 289.1706.

4.1.13   7-Allyl-2-trifluoromethyl-7,8-dihydro-6H-benzo[f]indolo[1,2-a][1,4]diazepine (3h)

By a procedure similar to that described for indole **3a**, **1e** (33.7 mg, 0.13 mmol) was converted into **3h** (23.1 mg, 53%) as an yellow oil. ^{1}H NMR (500 MHz, CDCl$_3$) δ 3.19 (d, J = 6.9 Hz, 2H, NCH$_2$CH), 3.29 (d, J = 12.6 Hz, 1H, NCHH), 3.44 (d, J = 13.9 Hz, 1H, NCHH), 3.68 (d, J = 12.6 Hz, 1H, NCHH), 3.95 (d, J = 13.9 Hz, 1H, NCHH), 5.26 (dd, J = 10.2, 1.7 Hz, 1H, CH=CHH), 5.31 (dd, J = 17.1, 1.6 Hz, 1H, CH=CHH), 5.93–6.01 (m, 1H, CH), 6.62 (s, 1H, 3-H), 7.38 (ddd, J = 7.4, 7.4, 1.4 Hz, 1H, Ar), 7.42 (dd, J = 8.3, 1.1 Hz, 1H, Ar), 7.46 (dd, J = 7.4, 1.4 Hz, 1H, Ar), 7.56 (ddd, J = 7.4, 7.4, 1.4 Hz, 1H, Ar), 7.67 (dd, J = 7.4, 1.4 Hz, 1H, Ar), 7.74 (d, J = 8.3 Hz, 1H, Ar), 7.87 (m, 1H, Ar); ^{13}C NMR (125 MHz, CDCl$_3$) δ 47.7, 54.2, 59.0, 102.4, 107.8 (d, J = 3.6 Hz), 117.3 (d, J = 3.6 Hz), 118.6, 121.2, 123.0, 124.4 (q, J = 32.4 Hz), 125.1 (q, J = 272.3 Hz), 126.9, 129.2, 130.5, 131.1, 131.4, 134.8, 135.6, 137.9, 138.6; MS

(FAB) m/z (%): 343 (MH$^+$, 25); HRMS (FAB) calcd for $C_{20}H_{18}F_3N_2$ (MH$^+$): 343.1422; found: 343.1427.

4.1.14 Synthesis of N-[(2-bromothiophen-3-yl)methyl]butan-1-amine (4)

To a stirred solution of 2-bromo-3-(bromomethyl)thiophene (1.90 g, 7.42 mmol) in EtOH (5 mL) was added dropwise n-BuNH$_2$ (7.40 mL, 74.23 mmol) at rt. The reaction mixture was stirred for 3 h at this temperature and extracted with EtOAc. The extract was washed with H$_2$O, dried over MgSO$_4$ and concentrated under reduced pressure. The residue was purified by column chromatography over silica gel with hex/AcOEt (3:1 to 1:1) to give 4 (1.63 g, 88%) as an yellow oil; ^{1}H NMR (500 MHz, CDCl$_3$) δ 0.91 (t, $J = 7.4$ Hz, 3H, CH$_3$), 1.31–1.38 (m, 2H, CH$_2$CH$_3$), 1.45–1.51 (m, 2H, CH$_2$CH$_2$CH$_3$), 2.61 (dd, $J = 7.4$, 7.4 Hz, 2H, CH$_2$CH$_2$CH$_2$CH$_3$), 3.73 (s, 2H, ArCH$_2$), 6.94 (d, $J = 5.7$ Hz, 1H, Ar), 7.21 (d, $J = 5.7$ Hz, 1H, Ar); ^{13}C NMR (125 MHz, CDCl$_3$) δ 13.9, 20.4, 32.1, 47.6, 49.0, 110.0, 125.6, 128.2, 140.3; MS (FAB) m/z (%): 248 [MH$^+$ (^{79}Br), 100] 250 [MH$^+$ (^{81}Br), 100]; HRMS (FAB) calcd for $C_{19}H_{15}BrNS$ (MH$^+$): 248.0109; found: 248.0100.

4.1.15 7-Allyl-7,8-dihydro-6H-pyrydo[3,2-f]indolo[1,2-a][1,4] diazepine (5)

By a procedure similar to that described for indole 3a, 1a (25.0 mg, 0.13 mmol) was converted into 5 (24.9 mg, 71%) as an yellow oil by treatment with 4; ^{1}H NMR (500 MHz, CDCl$_3$) δ 3.24 (d, $J = 6.6$ Hz, 2H, NCH$_2$CH), 3.52 (s, 2H, NCH$_2$), 3.76 (s, 2H, NCH$_2$), 5.27 (dd, $J = 10.2$, 1.4 Hz, 1H, CH=CHH), 5.32 (dd, $J = 17.0$, 1.4 Hz, 1H, CH=CHH), 5.94–6.02 (m, 1H, CH), 6.57 (s, 1H, 3-H), 7.19–7.29 (m, 3H, Ar), 7.63 (d, $J = 8.0$ Hz, 1H, Ar), 7.74 (dd, $J = 7.4$, 1.8 Hz, 1H, Ar), 8.09 (d, $J = 8.0$ Hz, 1H, Ar), 8.59 (dd, $J = 4.9$, 1.8 Hz, 1H, Ar); ^{13}C NMR (125 MHz, CDCl$_3$) δ 48.0, 54.1, 59.1, 104.3, 112.5, 118.7, 120.5, 120.8, 121.4, 123.0, 124.5, 128.8, 134.6, 135.4, 136.1, 139.8, 148.4, 152.7; MS (FAB) m/z (%): 276 (MH$^+$, 71); HRMS (FAB) calcd for $C_{18}H_{18}N_3$ (MH$^+$): 276.1501; found: 276.1507.

4.1.16 Synthesis of N-((2-bromopyridin-3-yl)methyl)prop-2-en-1-amine (6)

A mixture of 2-bromopicolinaldehyde (0.5 g, 0.27 mmol) and allylamine (2.00 mL, 2.72 mmol) in MeOH (1.5 mL) was stirred at rt for 48 h. To the mixture was added NaBH$_4$ (0.11 g, 0.29 mmol) at 0 °C and the mixture was stirred at this temperature for 10 min. After quenched with H$_2$O, the mixture was extracted with EtOAc and the organic layer was washed with H$_2$O, dried over MgSO$_4$, and concentrated under reduced pressure. The residue was purified by column chromatography over silica gel with hex/EtOAc (2:1 to 1:1) to give **6** (0.11 g, 35%) as an yellow oil; ^{1}H NMR (500 MHz, CDCl$_3$) δ 3.29 (d, $J = 5.7$ Hz, 2H, CH=CH$_2$), 3.85 (s, 2H, ArCH$_2$), 5.13–5.16 (m, 1H, CHH), 5.21–5.25 (m, 1H, CH=CHH), 5.89–5.97 (m, 1H, CH=CH$_2$), 7.26 (dd, $J = 7.4$, 4.6 Hz, 1H, Ar), 7.76 (dd, $J = 7.4$, 1.7 Hz, 1H, Ar), 8.26 (dd, $J = 4.6$, 1.7 Hz, 1H, Ar); ^{13}C NMR (125 MHz, CDCl$_3$) δ 51.6, 51.7, 116.4, 122.8, 136.3, 136.6, 138.0, 143.5, 148.3; MS (FAB) m/z (%): 227 [MH$^+$ (^{79}Br), 100], 229 [MH$^+$ (^{81}Br), 80]; HRMS (FAB) calcd for C$_9$H$_{12}$BrN$_2$ (MH$^+$): 227.0184; found: 227.0176.

4.1.17 7-Allyl-7,8-dihydro-6H-indolo[1,2-a]thieno[2,3-f][1,4] diazepine (7)

By a procedure similar to that described for indole **3a**, **1a** (25.0 mg, 0.13 mmol) was converted into **7** (21.2 mg, 56%) as an yellow oil by treatment with **6**. ^{1}H NMR (500 MHz, CDCl$_3$) δ 0.95 (t, $J = 7.4$ Hz, 3H, CH$_3$), 1.34–1.42 (m, 2H, CH$_2$CH$_3$), 1.54–1.60 (m, 2H, CH$_2$CH$_2$CH$_3$), 2.57 (dd, 2H, NCH$_2$CH$_2$), 3.55 (s, 2H, NCH$_2$), 3.74 (s, 2H, NCH$_2$), 6.57 (s, 1H, 3-H), 6.99 (d, $J = 5.3$ Hz, 1H, Ar), 7.11 (d, $J = 5.3$ Hz, 1H, Ar), 7.18–7.21 (m, 1H, Ar), 7.25–7.28 (m, 1H, Ar), 7.63 (d, $J = 8.0$ Hz, 1H, Ar), 7.80 (d, $J = 8.0$ Hz, 1H, Ar); ^{13}C NMR (125 MHz, CDCl$_3$) δ 14.1, 20.7, 30.2, 49.2, 50.5, 56.3, 103.2, 110.4, 119.2, 120.8, 121.1, 122.5, 127.5, 128.6, 129.2, 136.1, 137.2, 137.4; MS (FAB) m/z (%): 297 (MH$^+$, 100); HRMS (FAB) calcd for C$_{18}$H$_{20}$N$_2$NaS (MNa$^+$): 319.1245; found: 319.1261.

References

1. Wasilke J-C, Obrey SJ, Baker RT, Bazan GC (2005) Chem Rev 105:1001–1020
2. Nicolaou KC, Edmonds DJ, Bulger PG (2006) Angew Chem Int Ed 45:7134–7186
3. Burk MJ, Lee JR, Martinez JP (1994) J Am Chem Soc 116:10847–10848
4. Jeong N, Seo SD, Shin JY (2000) J Am Chem Soc 122:10220–10221
5. Bielawski CW, Louie J, Grubbs RH (2000) J Am Chem Soc 122:12872–12873
6. Son SU, Park KH, Seo H, Chung YK, Lee S-G (2001) Chem Commun 2440–2441
7. Sutton AE, Seigal BA, Finnegan DF, Snapper ML (2002) J Am Chem Soc 124:13390–13391
8. Komon ZJA, Diamond GM, Leclerc MK, Murphy V, Okazaki M, Bazan GC (2002) J Am Chem Soc 124:15280–15285
9. Dijk EW, Panella L, Pinho P, Naasz R, Meetsma A, Minnaard AJ, Feringa BL (2004) Tetrahedron 60:9687–threetwo9693
10. van As BAC, van Buijtenen J, Heise A, Broxterman QB, Verzijl GKM, Palmans ARA, Meijer EW (2005) J Am Chem Soc 127:9964–9965

11. Alexakis A, Benhaim C (2002) Eur J Org Chem 19:3221–3236
12. Ley SV, Thomas AW (2003) Angew Chem Int Ed 42:5400–5449
13. Chemler SR, Fuller PH (2007) Chem Soc Rev 36:1153–1160
14. Carril M, SanMartin R, Domínguez E (2008) Chem Soc Rev 37:639–647
15. Hiroya K, Itoh S, Ozawa M, Kanamori Y, Sakamoto T (2002) Tetrahedron Lett 43: 1277–1280
16. Kamijo S, Sasaki Y, Yamamoto Y (2004) Tetrahedron Lett 45:35–38
17. Li K, Alexakis A (2005) Tetrahedron Lett 46:8019–8022
18. Loones KTJ, Maes BUW, Meyers C, Deruytter J (2006) J Org Chem 71:260–264
19. Yuen J, Fang Y-Q, Lautens M (2006) Org Lett 8:653–656
20. Zhang L, Malinakova HC (2007) J Org Chem 72:1484–1487
21. Martin R, Laursen CH, Cuenca A, Buchwald SL (2007) Org Lett 9:3379–3382
22. Français A, Urban D, Beau J-M (2007) Angew Chem Int Ed 46:8662–8665
23. Kumaraswamy G, Ankamma K, Pitchaiah A (2007) J Org Chem 72:9822–9825
24. Maryanoff BE, Nortey SO, Gardocki JF (1984) J Med Chem 27:1067–1071
25. Ho CY, Hageman WE, Persico FJ (1986) J Med Chem 29:1118–1121
26. Suzuki H, Shinpo K, Yamazaki T, Niwa S, Yokoyama Y, Murakami Y (1996) Heterocycles 42:83–threetwo86
27. Sasaki S, Ehara T, Sakata I, Fujino Y, Harada N, Kimura J, Nakamura H, Maeda M (2001) Bioorg Med Chem Lett 11:583–threetwo585
28. Kau TR, Schroeder F, Ramaswamy S, Wojciechowski CL, Zhao JJ, Roberts TM, Clardy J, Sellers WR, Silver PA (2003) Cancer Cell 4:463–threetwo476
29. Ennis MD, Hoffman RL, Ghazal NB, Olson RM, Knauer CS, Chio CL, Hyslop DK, Campbell JE, Fitzgerald LW, Nichols NF, Svensson KA, McCall RB, Haber CL, Kagey ML, Dinh DM (2003) Bioorg Med Chem Lett 13:2369–threetwo2372
30. Ducker CE, Griffel LK, Smith RA, Keller SN, Zhuang Y, Xia Z, Diller JD, Smith CD (2006) Mol Cancer Ther 5:1647–threetwo1659
31. Yang S-M, Malaviya R, Wilson LJ, Argentieri R, Chen X, Yang C, Wang B, Cavender D, Murray WV (2007) Bioorg Med Chem Lett 17:326–331
32. Ohno H, Ohta Y, Oishi S, Fujii N (2007) Angew Chem Int Ed 46:2295–2298
33. Ohta Y, Oishi S, Fujii N, Ohno H (2008) Chem Commun 835–837
34. Ivashchenko AV, Ilyin AP, Kysil VM, Trifilenkov AS, Tsirulnikov SA, Shkirando AM, Churakova MV, Lomakina IO, Potapov VV, Zamaletdinova AI, Tkachenko SY, Kravchenko DV, Khvat AV, Okun IM, Kyselev AS (2007) PCT Int Appl WO2007117180
35. Antilla JC, Klapars A, Buchwald SL (2002) J Am Chem Soc 124:11684–11688
36. Kabalka GW, Wang L, Pagni RM (2001) Tetrahedron 57:8017–80128
37. Wang H, Jiang Y, Gao JK, Ma D (2009) Tetrahedron 65:8956–8960

Part II
Synthesis of Isoquinoline Derivatives

Chapter 5
Facile Synthesis of 3-(Aminomethyl)isoquinoline by Copper-Catalyzed Domino Four-Component Coupling and Cyclization

5.1 Introduction

In Chap. 2, the author has reported an efficient construction of 2-(amino-methyl)indoles by a copper-catalyzed three-component coupling–cyclization reaction [1, 2]. This reaction proceeds through Mannich-type coupling followed by indole formation. On the basis of this indole synthesis, the author expected that a four-component coupling reaction of 2-ethynylbenzaldehyde 1, aldehyde 2, secondary amine 3, and an appropriate N-1 synthon 4 followed by cyclization of the alkyne intermediate 5 having a nitrogen atom with proximity to the triple bond (for copper-catalyzed isoquinoline formation through N-tert-butyl-2-(1-alkynyl)benz-aldimine derivatives, see [3–7]; For other isoquinoline formation from related intermidiates, see [8–15]) would provide a direct route to 3-(aminomethyl)iso-quinolines 6 without wasting any salts (Scheme 1). In this Section, the author describes a copper-catalyzed domino four-component coupling–cyclization reaction for diversity-oriented synthesis of 3-(aminomethyl)isoquinolines. To the best of the author's knowledge, this is the first example of four-component synthesis of an isoquinoline scaffold. For synthesis of isoquinolines by three-component reaction, see [16, 17].

In the initial investigation, the author examined the effect of N-1 synthon on the copper-catalyzed four-component synthesis of 3-(aminomethyl)isoquinoline using 2-ethynyl benzaldehyde 1a as a model substrate, paraformaldehyde 2 and diiso-propylamine 3a (Table 1). Since two nucleophilic reagents coexist with two aldehydes in the reaction system, the nucleophilic reactions in the desired order might be hampered on one-potion reaction. Actually, one-portion addition of all the four components using 4j gave a complex mixture of unidentified products without producing 6 (compare with Table 1, entry 10). Accordingly, after the copper-catalyzed three-component reaction of 1a, 2, and 3a in DMF was completed (monitored by TLC), N-1 synthon was added. Whereas ammonium nitrite 4a, perchlorate 4b, hydroxide 4c, formate 4d, chloride 4e, and sulfate 4f were

Y. Ohta, *Copper-Catalyzed Multi-Component Reactions*, Springer Theses,
DOI: 10.1007/978-3-642-15473-7_5, © Springer-Verlag Berlin Heidelberg 2011

Scheme 1 Construction of 3-(aminomethyl)isoquinolines by copper-catalyzed four-component coupling–cyclization

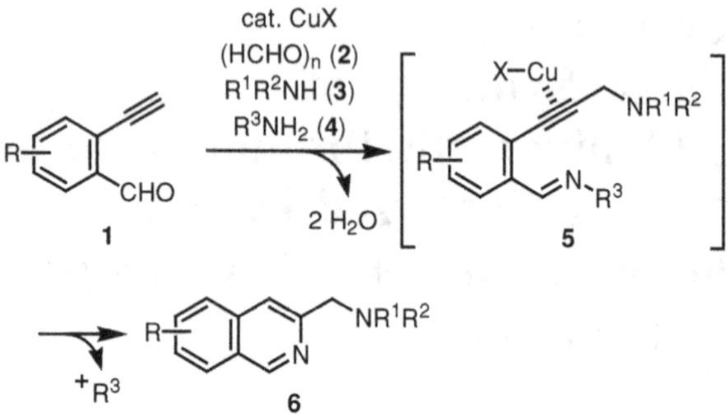

Table 1 Optimization of N-1 synthon **4**

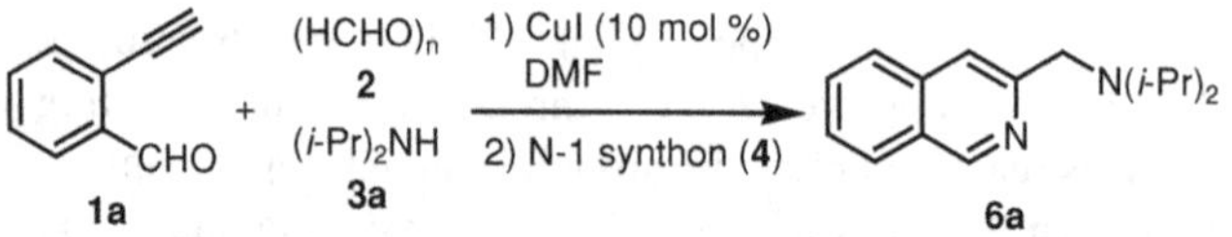

Entry	N-1 synthon	Yield (%)[a]
1	NH$_4$NO$_2$ (**4a**)	Decomp.
2	NH$_4$ClO$_4$ (**4b**)	Decomp.
3	28% NH$_4$OH (**4c**)	Trace
4	HCO$_2$NH$_4$ (**4d**)	Trace
5	NH$_4$Cl (**4e**)	Trace
6	(NH$_4$)$_2$SO$_4$ (**4f**)	Trace
7	AcONH$_4$ (**4g**)	42
8	NH$_4$HCO$_3$ (**4h**)	53
9	2,4,6-(MeO)$_3$C$_6$H$_2$CH$_2$NH$_2$·HCl (**4i**)	82
10	t-BuNH$_2$ (**4j**)	83

After a mixture of 2-ethynylbenzaldehyde **1a**, paraformaldehyde **2** (2 equiv), amine **3a** (2 equiv) and CuI (10 mol%) in DMF was stirred at rt for 1 h, and N-1 synthon **4** (6 equiv) was added. The resulting mixture was stirred for 5 h at rt and additional 45 min at 140 °C
[a] Isolated yield

ineffective (entries 1–6), the use of acetate **4g** and hydrogen carbonate **4h** gave, as expected, the desired isoquinoline **6a** in moderate yields (42–53%, entries 7 and 8). For isoquinoline formation with such ammonium salts as formate, carbonate, and ammonia, see Ref. [9]. More promising results were obtained with primary amines having a readily cleavable alkyl group such as 2,4,6-trimethoxybenzylamine hydrochloride **4i** and *tert*-butylamine **4j** [3–7], leading to high yield of **6a**. Taking the atom economy of the reaction into consideration, the author regarded **4j** as the most potent N-1 synthon.

Table 2 Synthesis of various 3-(aminomethyl)isoquinolines

Entry	Amine	Conditions[a]	Product	Yield (%)[c]
1	i-Pr$_2$NH **3a**	rt 1 h	**6a**	83
2	Bn$_2$NH **3b**	100 °C 1 h	**6b**	0
3	**3c**	100 °C 1 h	**6c**	73
4	(allyl)$_2$NH **3d**	rt 1 h[b]	**6d**	60
5	**3e**	rt 1 h[b]	**6e**	88
6	**3f**	rt 1 h[b]	**6f**	79

After the three-component reaction of **1a**, **2** (2 equiv), and **3** (2 equiv) in the presence of CuI (10 mol%) in DMF was completed (monitored by TLC), t-BuNH$_2$ **4j** (6 equiv) was added and the reaction mixture was stirred for 5 h at rt and additional 45 min at 140 °C
[a] Conditions for the three-component coupling
[b] Before **1a** was added, a mixture of **2**, **3** and CuI in DMF was stirred for 30 min at rt
[c] Isolated yield

Next, various secondary amines were employed to determine the scope of this reaction (Table 2). Although dibenzylamine **3b** showed lower reactivity toward Mannich-type coupling with **1a** and **2** leading to recovery of the unchanged starting material (entry 2), the reaction with more bulky bis(1-phenylethyl)amine **3c** led to successful conversion into the corresponding isoquinoline **6c** (73%, entry 3). Unfortunately, the initial Mannich-type reaction with highly nucleophilic diallylamine, piperidine, or pyrrolidine was unsuccessful producing a complex

Table 3 Reactions with various substituted 2-ethynylbenzaldehyde

Entry	Substrate	Product	Yield (%)[a]
1	1b	7	83
2	1c	8	79
3	1d	9	87
4	1e	10	84

After the three-component reaction of **1**, **2** (2 equiv), and **3a** (2 equiv) in the presence of CuI (10 mol%) in DMF was completed on TLC, t-BuNH$_2$ (**4j**, 6 equiv) was added and the reaction mixture was stirred for 5 h at rt and additional 45 min at 140 °C
[a] Isolated yield

mixture, presumably due to the simultaneous presence of two aldehydes (2-ethynylbenzaldehyde **1a** and paraformaldehyde **2**) and a reactive amine. Extensive optimization of the reaction conditions revealed that the addition of 2-ethynylaldehyde **1a** after the formation of iminium between secondary amines **3d–f** and paraformaldehyde **2** effectively produced 3-(aminomethyl)isoquinolines **6d–f**, respectively, in moderate to high yields (entries 4–6).

The copper-catalyzed domino four-component synthesis of 3-(amino-methyl)isoquinolines with various substituted 2-ethynylbenzaldehyde was next investigated (Table 3). The use of 2-ethynyl-4-fluorobenzaldehyde **1b** in the presence of CuI (10 mol%) gave the desired 3-(aminomethyl)-6-fluoroisoquinoline derivative **7** in high yield (83%, entry 1). Benzaldehyde **1c** which has a fluorine atom at the *meta*-position to the formyl group afforded the corresponding isoquinoline **8** (79%, entry 2). Also in the case of 2-ethynylbenzaldhydes containing an electron-donating group such as methyl or methoxy group at the *para*- or *meta*-position to the formyl group (**1d** and **1e**, respectively), the copper-catalyzed four-component isoquinoline formation proceeded smoothly (87 and 84% yield, respectively, entries 3 and 4). Thus, this isoquinoline formation has proven to be widely applicable to 2-ethynylbenzaldehydes having an electron-withdrawing and -donating group.

In conclusion, the author has developed a novel copper-catalyzed domino four-component coupling–cyclization reaction for the synthesis of 3-(amino-methyl)isoquinolines, which form one carbon–carbon and three carbon–nitrogen bonds. This methodology could be applied to construction of a highly potent isoquinoline library in terms of diversity and biological activity.

5.2 Experimental Section

5.2.1 General Methods

IR spectra were determined on a JASCO FT/IR-4100 spectrometer. Exact mass (HRMS) spectra were recorded on JMS-HX/HX 110A mass spectrometer. ^{1}H NMR spectra were recorded using a JEOL AL-400 spectrometer at 400 MHz frequency. Chemical shifts are reported in δ (ppm) relative to Me$_4$Si (in CDCl$_3$) as internal standard. ^{13}C NMR spectra were recorded using a JEOL AL-400 and referenced to the residual CHCl$_3$ signal. Melting points (uncorrected) were measured by a hot stage melting point apparatus. For column chromatography, Wakosil C-300 was employed.

5.2.1.1 2-Ethynylbenzaldehyde (1a)

To a stirred suspension of 2-bromobenzaldehyde (2.00 g, 10.8 mmol), PdCl$_2$-(PPh$_3$)$_2$ (152 mg, 0.22 mmol), CuI (41.2 mg, 0.22 mmol) was added trimethylsilylacetylene (1.77 mL, 12.97 mmol) at rt under argon. The reaction mixture was stirred for 30 min at 80 °C followed by filtration though a pad of Celite. The filtrate was concentrated under reduced pressure and the residue was purified by column chromatography over silica gel with hexane/AcOEt (50:1) as the eluent to give a solid mass. This solid was treated with K$_2$CO$_3$ (0.50 g, 3.64 mmol) in MeOH (20 mL) for 15 min at rt, and the solvent was removed under the reduced pressure. The residue was extracted with CH$_2$Cl$_2$ and the extract was washed with saturated aqueous Na$_2$CO$_3$, and dried over MgSO$_4$. The filtrate was concentrated under reduced pressure to give a solid mass which was purified by column chromatography over silica gel with hexane/AcOEt (50:1) to give the title compound 1a (0.87 g, 62% yield from 2-bromobenzaldehyde). Recrystallization from n-hexane gave pure 1a as colorless crystals: mp 65 °C; IR (neat): 2097 (C≡C), 1686 (C=O); ^{1}H NMR (400 MHz, CDCl$_3$) δ 3.47 (s, 1H, C≡CH), 7.47–7.52 (m, 1H, Ar), 7.55–7.63 (m, 2H, Ar), 7.92–7.95 (m, 1H, Ar), 10.54 (d, J = 0.7 Hz, 1H, CHO); ^{13}C NMR (100 MHz, CDCl$_3$) δ 79.2, 84.2, 125.5, 127.2, 139.2, 133.7, 133.9, 136.6, 191.4. Anal. Calcd for C$_9$H$_6$O: C, 83.06; H, 4.65. Found: C, 82.99; H, 4.61.

5.2.1.2 2-Ethynyl-4-fluorobenzaldehyde (1b)

By a procedure identical to that described for 1a, 2-bromo-4-fluorobenzaldehyde (1.00 g, 4.93 mmol) was converted to 1b (434 mg, 62%) as a solid mass, which was recrystallized from n-hexane: colorless crystals; mp 103 °C; IR (neat): 2103 (C≡C), 1689 (C=O); ^{1}H NMR (400 MHz, CDCl$_3$) δ 3.52 (s, 1H, C≡CH), 7.16–7.21 (m, 1H, Ar), 7.29 (dd, $J = 8.8$, 2.4 Hz, 1H, Ar), 7.97 (dd, $J = 8.8$, 5.9 Hz, 1H, Ar), 10.46 (s, 1H, CHO); ^{13}C NMR (100 MHz, CDCl$_3$) δ 78.0, 85.4, 117.2 (d, $J = 21.6$ Hz), 120.5 (d, $J = 24.0$ Hz), 127.9 (d, $J = 10.8$ Hz), 130.1 (d, $J = 9.6$ Hz), 133.3 (d, $J = 3.6$ Hz), 165.5 (d, $J = 256.3$ Hz), 189.7. Anal. Calcd for C$_9$H$_5$FO: C, 72.97; H, 3.40. Found: C, 73.09; H, 3.14.

5.2.1.3 2-Ethynyl-5-fluorobenzaldehyde (1c)

By a procedure identical to that described for **1a**, 2-bromo-5-fluorobenzaldehyde (1.00 g, 4.93 mmol) was converted to **1c** (542 mg, 77%) as a solid mass which was recrystallized from n-hexane: colorless crystals; mp 109 °C; IR (neat): 2101 (C≡C), 1693 (C=O); ^{1}H NMR (400 MHz, CDCl$_3$) δ 3.46 (s, 1H, C≡CH), 7.25–7.30 (m, 1H, Ar), 7.59–7.64 (m, 2H, Ar), 10.49 (d, $J = 3.2$ Hz, 1H, CHO); ^{13}C NMR (100 MHz, CDCl$_3$) δ 78.2, 84.0, 113.8 (d, $J = 22.8$ Hz), 121.2 (d, $J = 22.8$ Hz), 121.5 (d, $J = 3.6$ Hz), 135.9 (d, $J = 7.2$ Hz), 138.6 (d, $J = 7.2$ Hz), 162.7 (d, $J = 254.3$ Hz), 190.1. Anal. Calcd for C$_9$H$_5$FO: C, 72.97; H, 3.40. Found: C, 73.26; H, 3.31.

5.2.1.4 2-Ethynyl-4-methylbenzaldehyde (1d)

By a procedure identical with that described for **1a**, 2-bromo-4-methylbenzalde-hyde (1.00 g, 5.02 mmol) was converted to **1d** (555 mg, 76%) as a solid mass which was recrystallized from n-hexane: colorless crystals; mp 81 °C; IR (neat): 2101 (C≡C), 1685 (C=O); ^{1}H NMR (400 MHz, CDCl$_3$) δ 2.41 (s, 3H, CH$_3$), 3.42 (s, 1H, C≡CH), 7.29 (d, $J = 8.0$ Hz, 1H, Ar), 7.43 (s, 1H, Ar), 7.83 (d, $J = 8.0$ Hz, 1H, Ar), 10.47 (s, 1H, CHO); ^{13}C NMR (100 MHz, CDCl$_3$) δ 21.5, 79.4, 83.7, 125.5, 127.3, 130.2, 134.3, 134.4, 144.8, 191.1. Anal. Calcd for C$_{10}$H$_8$O: C, 83.31; H, 5.59. Found: C, 83.29; H, 5.74.

5.2.1.5 2-Ethynyl-5-methoxybenzaldehyde (1e)

By a procedure identical with that described for **1a**, 2-bromo-5-methoxybenzal-dehyde (1.00 g, 4.65 mmol) was converted to **1e** (593 mg, 80%) as a solid mass which was recrystallized from n-hexane: colorless crystals; mp 98 °C; IR (neat): 2098 (C≡C), 1677 (C=O); ^{1}H NMR (400 MHz, CDCl$_3$) δ 3.37 (s, 1H, C≡CH), 3.87 (s, 3H, OMe), 7.11 (dd, $J = 8.5$, 2.7 Hz, 1H, Ar), 7.41 (d, $J = 2.7$ Hz, 1H, Ar), 7.53 (d, $J = 8.5$ Hz, 1H, Ar), 10.50 (s, 1H, CHO); ^{13}C NMR (100 MHz, CDCl$_3$) δ 55.6, 79.2, 82.7, 109.9, 118.1, 121.4, 135.2, 138.0, 160.1, 191.3. Anal. calcd for C$_{10}$H$_8$O$_2$: C, 74.99; H, 5.03. Found: C, 75.15; H, 4.81.

5.2.2 General Procedure for Four-Component Isoquioline Formation

5.2.2.1 Synthesis of 3-[(Diisopropylaminino)methyl]isoquinoline (6a)

To a stirred suspension of 2-ethynylbenzaldehyde 1a (25 mg, 0.19 mmol), $(HCHO)_n$ 2 (12 mg, 0.38 mmol), and CuI (3.7 mg, 0.019 mmol) in DMF (1.5 mL) was added i-Pr$_2$NH 3a (54 μL, 0.38 mmol) at rt under Ar. After the reaction mixture was stirred for 1 h at this temperature, t-BuNH$_2$ 4j (121 μL, 1.2 mmol) was added and the mixture was stirred for 6 h at rt before stirring for 45 min at 140 °C. The reaction mixture was concentrated in vacuo and purified by column chromatography over alumina with hexane/AcOEt (50:1) as the eluent to give 6a (38.6 mg, 83% yield) as a pale yellow oil: ^{1}H NMR (400 MHz, CDCl$_3$) δ 1.08 (d, J = 6.6 Hz, 12H, 4 × CH$_3$), 3.09–3.19 (m, 2H, 2 × NCH), 3.97 (s, 2H, NCH$_2$), 7.48–7.52 (m, 1H, Ar), 7.61–7.65 (m, 1H, Ar), 7.80 (d, J = 7.6 Hz, 1H, Ar), 7.91–7.93 (m, 1H, Ar, 4-H), 9.16 (s, 1H, 1-H); ^{13}C NMR (100 MHz, CDCl$_3$) δ 20.8 (4C), 49.1 (2C), 51.3, 117.4, 126.1, 126.5, 127.4, 127.5, 130.0, 136.6, 151.5, 157.4; MS (FAB) m/z (%): 243 (MH$^+$, 100); HRMS (FAB) calcd for C$_{16}$H$_{23}$N$_2$ (MH$^+$): 243.1861; found: 243.1857.

5.2.2.2 3-{Bis[(R)-1-phenylethyl]aminomethyl}isoquinoline (6c)

To a stirred suspension of 2-ethynylbenzaldehyde 1a (25 mg, 0.19 mmol), $(HCHO)_n$ 2 (12 mg, 0.38 mmol), and CuI (3.7 mg, 0.019 mmol) in DMF (1.5 mL) was added (+)-bis[(R)-1-phenylethyl]amine 6c (87.8 μL, 0.38 mmol) at rt under Ar. After the reaction mixture was stirred for 1 h at 100 °C followed by cooling to rt, t-BuNH$_2$ 4j (121 μL, 1.2 mmol) was added and the mixture was stirred for 6 h at rt before stirring for 45 min at 140 °C. The reaction mixture was concentrated in vacuo and purified by column chromatography over silica gel with hexane/AcOEt (7:1) as the eluent to give the desired product 6c (51.2 mg, 73% yield) as a pale yellow oil: ^{1}H NMR (400 MHz, CDCl$_3$): δ 1.34 (d, J = 6.9 Hz, 6H, 2 × CH$_3$), 3.84 (d, J = 16.6 Hz, 1H, NCH$_2$), 4.05 (q, J = 6.9 Hz, 2H, 2 × NCH), 4.42 (d, J = 16.6 Hz, 1H, NCH$_2$) 7.21–7.39 (m, 10H, Ar), 7.50–7.53 (m, 1H, Ar), 7.64–7.70 (m, 1H, Ar), 7.84 (d, J = 8.2, 1H, Ar), 7.92 (d, J = 8.2 Hz, 1H, Ar), 8.03 (s, 1H, 4-H), 9.12 (s, 1H, 1-H); ^{13}C NMR (100 MHz, CDCl$_3$): δ 20.3 (2C), 52.1, 59.1 (2C), 118.0, 126.3, 126.5, 126.7 (2C), 127.5, 127.5, 127.8 (4C), 128.1 (4C), 130.1, 136.5, 144.2 (2C), 151.4, 157.3; MS (FAB) m/z (%): 367 (MH$^+$, 60); HRMS (FAB) calcd for C$_{26}$H$_{27}$N$_2$ (MH$^+$): 367.2174; found: 367.2169.

5.2.2.3 3-[(Diallylamino)methyl]isoquinoline (6d)

After the mixture of $(HCHO)_n$ 2 (12 mg, 0.38 mmol), and diallylamine 3d (47.4 μL, 0.38 mmol) and CuI (3.7 mg, 0.019 mmol) in DMF (1.5 mL) was stirred for 30 min at rt, 2-ethynylbenzaldehyde **1a** (25 mg, 0.19 mmol) was added and the reaction mixture was stirred for 1 h at rt. Then, *t*-BuNH$_2$ **4j** (121 μL, 1.2 mmol) was added and the reaction mixture was stirred for 6 h followed by being stirred for 45 min at 140 °C. The reaction mixture was concentrated in vacuo and purified by column chromatography over silica gel with CHCl$_3$/CH$_3$OH (50:1) as the eluent to give the desired product **6d** (27.4 mg, 60%) as a pale yellow oil: ^{1}H NMR (400 MHz, CDCl$_3$): δ 3.22 (d, $J = 6.3$ Hz, 4H, 2 × CH$_2$), 3.91 (s, 2H, CH$_2$), 5.17 (d, $J = 10.0$ Hz, 2H, CH=C*H*$_2$), 5.23 (d, $J = 17.3$ Hz, 2H, CH=C*H*$_2$), 5.96 (ddt, $J = 17.1$, 10.2, 6.3 Hz, 2H, 2 × C*H*=CH$_2$), 7.53–7.55 (m, 1H, Ar), 7.64–7.68 (m, 1H, Ar), 7.77 (s, 1H, 4-H). 7.80 (d, $J = 8.3$ Hz, 1H, Ar), 7.95 (d, $J = 8.3$ Hz, 1H, Ar), 9.22 (s, 1H, 1-H); ^{13}C NMR (100 MHz, CDCl$_3$): δ 56.9 (2C), 59.3, 117.7 (2C), 118.8, 126.5, 126.7, 127.5, 127.7, 130.3, 135.7 (2C), 136.4, 152.1, 153.1; MS (FAB) *m/z* (%): 239 (MH$^+$, 100); HRMS (FAB) calcd for C$_{16}$H$_{19}$N$_2$ (MH$^+$): 239.1548; found: 239.1554.

5.2.2.4 3-(Piperidin-1-ylmethyl)isoquinoline (6e)

By a procedure identical with that described for compound **6d** from the compound **1a**, **1a** (25 mg, 0.19 mmol) was converted to the compound **6e** (38.2 mg, 88%) as pale yellow oil: ^{1}H NMR (400 MHz, CDCl$_3$): δ 1.43–1.49 (m, 2H, CH$_2$), 1.61–1.67 (m, 4H, 2 × CH$_2$), 2.46–2.58 (m, 4H, 2 × NCH$_2$), 3.79 (s, 2H, NCH$_2$), 7.54–7.57 (m, 1H, Ar), 7.64–7.68 (m, 1H, Ar), 7.71 (s, 1H, 4-H), 7.80 (d, $J = 8.3$ Hz, 1H, Ar), 7.95 (d, $J = 8.3$ Hz, 1H, Ar), 9.23 (s, 1H, 1-H); ^{13}C NMR (100 MHz, CDCl$_3$): δ 24.4, 26.0 (2C), 54.9 (2C), 65.3, 119.2, 126.5, 126.7, 127.5, 127.9, 130.3, 136.3, 152.11, 152.14; MS (FAB) *m/z* (%): 227 (MH$^+$, 100); HRMS (FAB) calcd for C$_{15}$H$_{19}$N$_2$ (MH$^+$): 227.1548; found: 227.1552.

5.2.2.5 3-[(Pyrrolidin-1-yl)methyl]isoquinoline (6f)

By a procedure identical with that described for compound **6d** from the compound **1a**, the compound **1a** (25 mg, 0.19 mmol) was converted to the compound **6f** (32.4 mg, 79%) as a pale yellow oil: ^{1}H NMR (400 MHz, CDCl$_3$): δ 1.82–1.90 (m, 4H, 2 × CH$_2$), 2.70–2.76 (m, 4H, 2 × NCH$_2$), 4.00 (s, 2H, NCH$_2$), 7.56–7.59 (m, 1H, Ar), 766–7.69 (m, 1H, Ar), 7.76 (s, 1H, 4-H), 7.81 (d, $J = 8.3$ Hz, 1H, Ar), 7.96 (d, $J = 8.3$ Hz, 1H, Ar), 9.23 (s, 1H, 1-H); ^{13}C NMR (100 MHz, CDCl$_3$): δ 23.5 (2C), 54.2 (2C), 61.8, 119.2, 116.5, 126.9, 127.5, 127.7, 130.3, 136.3, 151.7, 152.1; MS (FAB) *m/z* (%): 213 (MH$^+$, 100); HRMS (FAB) calcd for C$_{14}$H$_{17}$N$_2$ (MH$^+$): 213.1392; found: 213.1396.

5.2.2.6 3-[(Diisopropylamino)methyl]-6-fluoroisoquinoline (7)

By a procedure identical with that described for compound **6a** from the compound **1a**, the compound **1a** (25 mg, 0.19 mmol) was converted to the compound **7**

(43.5 mg, 83%) as a pale yellow oil: ^{1}H NMR (400 MHz, CDCl$_3$): δ 1.08 (d, J = 6.6 Hz, 12H, 4 × CH$_3$), 3.14 (m, 2H, 2 × NCH), 3.95 (s, 2H, NCH$_2$), 7.25–7.30 (m, 1H, Ar), 7.41 (dd, J = 2.4, 9.8 Hz, 1H, Ar), 7.90–7.96 (m, 2H, 4-H, Ar), 9.12 (s, 1H, 1-H); ^{13}C NMR (100 MHz, CDCl$_3$): δ 20.8 (4C), 49.2 (2C), 51.3, 109.7 (d, J = 20.7 Hz), 116.7 (d, J = 26.5 Hz), 117.0 (d, J = 5.8 Hz), 124.7, 130.4 (d, J = 9.9 Hz), 138.1 (d, J = 10.8 Hz), 151.1, 158.6, 163.2 (d, J = 251.6 Hz); MS (FAB) m/z (%): 261 (MH$^+$, 100); HRMS (FAB) calcd for C$_{16}$H$_{22}$FN$_2$ (MH$^+$): 261.1767; found: 261.1764.

5.2.2.7 3-[(Diisopropylamino)methyl]-7-fluoroisoquinoline (8)

By a procedure identical with that described for compound **6a** from the compound **1a**, the compound **1a** (25 mg, 0.19 mmol) was converted to the compound **8** (41.7 mg, 79%) as a pale yellow oil: ^{1}H NMR (400 MHz, CDCl$_3$): δ 1.08 (d, J = 6.6 Hz, 12H, 4 × CH$_3$), 3.14 (m, 2H, 2 × NCH), 3.96 (s, 2H, NCH$_2$), 7.40–7.45 (m, 1H, Ar), 7.53 (dd, J = 8.8, 2.2 Hz, 1H, Ar), 7.81 (dd, J = 9.0, 5.4 Hz, 1H, Ar), 7.94 (s, 1H, 4-H), 9.12 (s, 1H, 1-H); ^{13}C NMR (100 MHz, CDCl$_3$): δ 20.8 (4C), 49.1 (2C), 51.2, 110.3 (d, J = 19.9 Hz), 117.3, 120.7 (d, J = 25.7 Hz), 127.8 (d, J = 8.3 Hz), 129.1 (d, J = 8.3 Hz), 133.7, 150.6 (d, J = 5.8 Hz), 157.1, 160.3 (d, J = 249.1 Hz); MS (FAB) m/z (%): 261 (MH$^+$, 100); HRMS (FAB) calcd for C$_{16}$H$_{22}$FN$_2$ (MH$^+$): 261.1767; found: 261.1766.

5.2.2.8 3-[(Diisopropylamino)methyl]-6-methylquinoline (9)

By a procedure identical with that described for compound **6a** from the compound **1a**, the compound **1a** (25 mg, 0.19 mmol) was converted to the compound **9** (38.9 mg, 87%) as an yellow oil: ^{1}H NMR (400 MHz, CDCl$_3$): δ 1.08 (d, J = 6.3 Hz, 12H, 4 × CH$_3$), 2.53 (s, 3H, CH$_3$), 3.14 (m, 2H, 2 × NCH), 3.94 (s, 2H, NCH$_2$), 7.34 (d, J = 8.3, 1.5 Hz, 1H, Ar), 7.57 (s, 1H, Ar), 7.82 (d, J = 8.3 Hz, 1H, Ar), 7.85 (s, 1H, 4-H), 9.09 (s, 1H, 1-H); ^{13}C NMR (100 MHz, CDCl$_3$): δ 20.8 (4C), 49.2 (2C), 51.3, 109.7, 116.7, 117.0, 124.7, 130.4, 138.1, 151.1, 158.6, 163.2; MS (FAB) m/z (%): 257 (MH$^+$, 100); HRMS (FAB) calcd for C$_{17}$H$_{25}$N$_2$ (MH$^+$): 257.2018; found: 257.2019.

5.2.2.9 3-[(Diisopropylamino)methyl]-7-methoxyquinoline (10)

By a procedure identical with that described for compound **6a** from the compound **1a**, the compound **1a** (25 mg, 0.19 mmol) was converted to the compound **10** (44.1 mg, 84%) as a pale yellow oil: ^{1}H NMR (400 MHz, CDCl$_3$): δ 1.08 (d, J = 6.3 Hz, 12H, 4 × CH$_3$), 3.13 (m, 2 H, 2 × NCH), 3.93 (s, 3H, OCH$_3$), 3.94 (s, 2H, NCH$_2$), 7.19 (d, J = 2.4 Hz, 1H, Ar), 7.31 (dd, J = 2.4 9.0 Hz, 1H, Ar), 7.71 (d, J = 9.0 Hz, 1H, Ar), 7.86 (s, 1H, 4-H), 9.07 (s, 1H, 1-H); ^{13}C NMR

(100 MHz, CDCl$_3$): δ 20.8 (4C), 49.0 (2C), 51.1, 55.4, 104.5, 117.3, 123.3, 128.0, 128.4, 132.4, 149.9, 155.5, 157.7; MS (FAB) m/z (%): 273 (MH$^+$, 100); HRMS (FAB) calcd for C$_{17}$H$_{25}$N$_2$O (MH$^+$): 273.1967; found: 273.1964.

References

1. Ohno H, Ohta Y, Oishi S, Fujii F (2007) Angew Chem Int Ed 46:2295–2298
2. Ohta Y, Chiba H, Oishi S, Fujii N, Ohno H (2009) J Org Chem 74:7052–7058
3. Roesh KR, Larock RC (1998) J Org Chem 63:5306–5307
4. Huang Q, Hunter JA, Larock RC (2001) Org Lett 3:2973–2976
5. Roesh KR, Larock RC (2002) J Org Chem 67:86–94
6. Huang Q, Hunter JA, Larock RC (2002) J Org Chem 67:3437–3444
7. Zhang H, Larock RC (2002) Tetrahedron Lett 43:1359–1362
8. Anderson PN, Sharp JT (1980) J Chem Soc Perkin Trans 1:1331–1334
9. Sakamoto T, Kondo Y, Miura N, Hayashi K, Yamanaka H (1986) Heterocycles 24:2311–2314
10. Sakamoto T, Numata A, Kondo Y (2000) Chem Pharm Bull 48:669–772
11. Dai G, Larock RC (2001) Org Lett 3:4035–4038
12. Huang Q, Larock RC (2002) Tetrahedron Lett 43:3557–3560
13. Asao N, Yudha SS, Nogami T, Yamamoto Y (2005) Angew Chem Int Ed 44:5526–5528
14. Yanada R, Obika S, Kono H, Takemoto Y (2006) Angew Chem Int Ed 45:3822–3825
15. Obika S, Kono H, Yasui Y, Yanada R, Takemoto Y (2007) J Org Chem 72:4462–4468
16. Asao N, Iso K, Yudha SS (2006) Org Lett 8:4149–4151
17. Oikawa M, Takeda Y, Naito S, Hashizume D, Koshino H, Sasaki M (2007) Tetrahedron Lett 48:4255–4258

Chapter 6
Rapid Access to 3-(Aminomethyl)isoquinoline-Fused Polycyclic Compounds by Copper-Catalyzed Four Component Coupling, Cascade Cyclization, and Oxidation

Isoquinoline-fused polycyclic compounds such as pyrimido[2,1-*a*]isoquinolines and imidazo[2,1-*a*]isoquinolines exert various biological effects [1–4] including anti-tumor activity [5–8]. Considerable efforts have been made to develop efficient methods for the synthesis of this class of compounds, in which stepwise introduction/construction of the desired ring system is generally required [9–18]. In Chap. 1, the author reported a novel synthesis of 3-(aminomethyl)isoquinolines by four-component coupling–cyclization (Scheme 1) [19]. In this reaction, a copper-catalyzed Mannich-type reaction of a 2-ethynylbenzaldehyde **1** with paraformaldehyde **2** and a secondary amine **3** followed by imine formation with *t*-BuNH₂ **4** promotes isoquinoline formation to afford **7** through cleavage of a *tert*-butyl group.

On the basis of this chemistry, the author expected that the use of a primary amine containing a tethered nucleophilic group instead of *t*-BuNH₂ could bring about an intramolecular nucleophilic attack onto the isoquinolinium ion **10** without causing cleavage (Scheme 2) [20–30]. In this section, the author describes a novel approach to 3-(aminomethyl)isoquinoline-fused polycyclic compounds utilizing four-component coupling and cascade cyclization in the presence of a copper catalyst. To the best of the author's knowledge, this is the first example of multi-component sequential construction of an isoquinoline-fused heterocyclic ring system including and pyrimido[2,1-*a*]isoquinolines.

The author envisioned that 1,3-diaminopropane would be an appropriate primary amine as it has an additional nucleophilic group that could sequentially form isoquinoline and pyrimidine rings (the reaction using 3-aminopropanol as the amine component **8** showed a promising result. However, the main product of this reaction was unstable and decomposed during purification). Thus, attempts to construct the pyrimido[2,1-*a*]isoquinoline framework was initiated with 2-ethynylbenzaldehyde **1a**, paraformaldehyde **2**, diisopropylamine **3a** and 1,3-diamino-propane **8a** (Table 1). Co-existence of two amines with two aldehydes in one-portion of the reaction would hamper the effective Mannich-type reaction of **1a**, **2** and **3a** and subsequent imine formation with **8a** in the desired order. Therefore, the copper-catalyzed Mannich-type reaction of **1a**, **2** (2 equiv) and **3a** (2

Y. Ohta, *Copper-Catalyzed Multi-Component Reactions*, Springer Theses, DOI: 10.1007/978-3-642-15473-7_6, © Springer-Verlag Berlin Heidelberg 2011

Scheme 1 Four-component synthesis of 3-(Amino-methyl)isoquinoline using copper catalysis

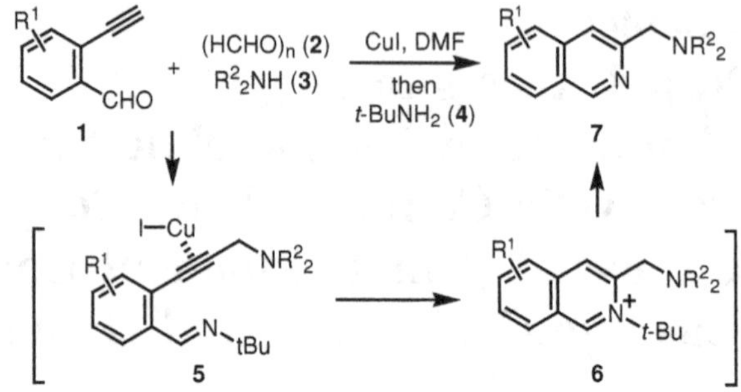

equiv) in DMF was completed (monitored by TLC), then the reaction mixture was treated with **8a** (3 equiv) at 120 °C to afford the expected product of the oxidized form **12a** in 38% yield (entry 1) (The unambiguous structure assignment for **12a** was made by X-ray analysis).

The elevated reaction temperature (200 °C) under microwave irradiation in the ring formation step led to a lower yield of **12a** (29%, entry 2). When other copper

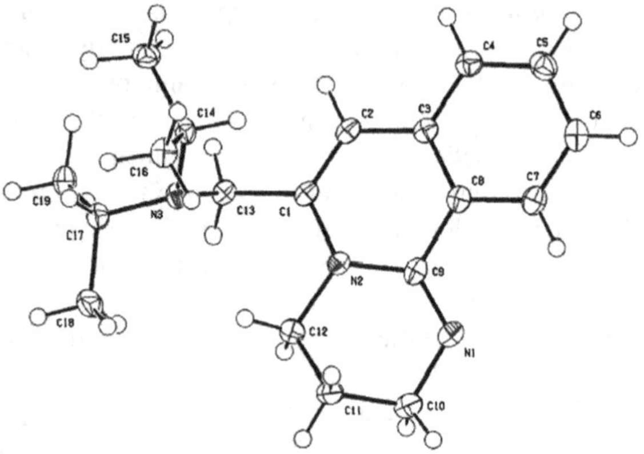

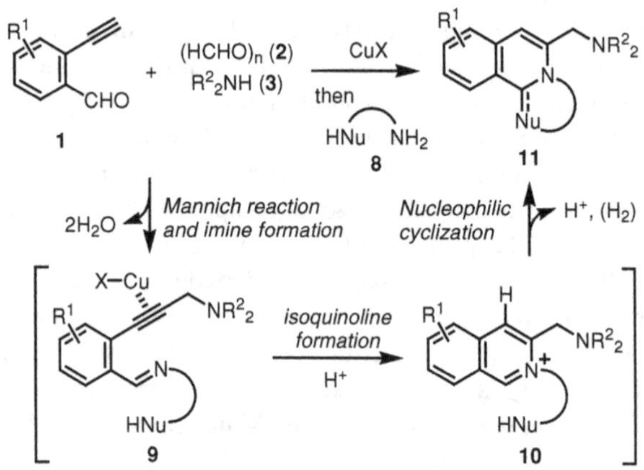

Scheme 2 Four-component construction of an isoquino-line-fused tricyclic ring system

salts such as CuBr, CuBr$_2$, CuCl$_2$, CuF$_2$, Cu(OAc)$_2$ and CuCl (entries 3–8) were used in the reaction, it was revealed that CuCl was the most effective catalyst for this transformation (43% yield, entry 8). Use of MS 4 Å slightly improved the yield of **12a** (52%, entry 9). Further optimization demonstrated that the cyclization reaction under an oxygen atmosphere, which would facilitate the oxidation step, realized rapid formation of **12a** in 72% yield (entry 10).

Several substituted 2-ethynylbenzaldehydes were then applied to this copper-catalyzed four-component synthesis of 3,4-dihydro-2*H*-pyrimido[2,1-*a*]isoquinoline under optimized conditions (Table 1, entry 10). The results are summarized in Table 2. The substitution by a fluorine atom at the *para*-position to the formyl group slightly decreased the yield of **12b** (55%, entry 1). The reaction with 2-ethynylbenzaldehydes **1c** and **1d** containing a fluorine atom at the *meta*-position or methyl group at the *para*-position to the formyl group showed a good conversion to yield the desired tricyclic compounds **12c** and **12d** (74 and 71%, respectively, entries 2, 3). The use of 2-ethynyl-5-methoxybenzaldehyde **1e** also gave tricyclic compound **12e** (55%, entry 4). Overall, this four-component construction of 3,

Table 1 Optimization of reaction conditions using 1,3-diaminopropane

Entry	CuX	Condition A	Condition B	Yield (%)[c]
1	CuI	rt, 0.5 h	120 °C, 15 h	38
2	CuI	rt, 0.5 h	MW, 200 °C, 0.33 h	29
3	CuBr	rt, 1.5 h	120 °C, 15 h	42
4	CuBr$_2$	rt, 1.0 h	120 °C, 15 h	38
5	CuCl$_2$	rt, 2.3 h	120 °C, 10 h	42
6	CuF$_2$	100 °C, 0.5 h	120 °C, 16 h	27
7	Cu(OAc)$_2$	rt, 2.5 h	120 °C, 12 h	20
8	CuCl	rt, 1.5 h	120 °C, 12 h	43
9[a]	CuCl	rt, 1.5 h	120 °C, 20 h	52
10[a, b]	CuCl	rt, 1.5 h	120 °C, 1 h	72

After the Mannich-type reaction of **1a**, **2** (2 equiv) and **3a** (2 equiv) in the presence of copper salt (10 mol %) was completed under conditions A (monitored by TLC), **8a** (3 equiv) was added. The reaction mixture was stirred under conditions B
[a] **8a** with MS 4 Å was added,
[b] Under oxygen atmosphere,
[c] Isolated yields

Table 2 Reaction with substituted 2-ethynylbenzaldehydes

Entry	2-ethynylbenzaldehyde	Product (yield)[a]
1	**1b**	**12b** (55%)
2	**1c**	**12c** (74%)
3	**1d**	**12d** (71%)
4	**1e**	**12e** (55%)

After the Mannich-type reaction of 1, 2 (2 equiv) and **3aa** (2 equiv) in the presence of CuCl (10 mol %) in DMF under O_2 was completed (rt, within 1.5 h, monitored by TLC), **8a** (2 equiv) and MS 4 Å were added and the reaction mixture was stirred at 120 °C for 1 h.
[a] Isolated yields

4-dihydro-2H-pyrimido[2,1-a]isoquinoline having an aminomethyl group was found to be applicable to 2-ethynylbenzaldehydes containing an electron-donating or electron-withdrawing group.

Next, investigation with several secondary amines **3** was conducted (Table 3). A one-portion Mannich-type reaction with 2-ethynylbenzaldehyde **1a**, paraformaldehyde **2** and piperidine **3b** was very sluggish. Therefore, a mixture of **2** and **3b** in DMF was allowed to react at rt for 1 h in the presence of CuCl before successive addition of **1a** and 1,3-diaminopropane **8a**. This stepwise addition was successful to give the desired 3,4-dihydro-2H-pyrimido[2,1-a]isoquinoline **12f** in 61% yield (entry 1). Diallylamine **3c** and bis(1-phenylethyl)amine **3d** showed

Table 3 Reaction with secondary amines **3b–d**

Entry	Secondary amine	Product (yield)[a]
1[b]	**3b**	**12f** (61%)
2[b]	allyl$_2$NH **3c**	N(allyl)$_2$ **12g** (30%)
3[c]	Ph⌣N(H)⌣Ph **3d**	**12h** (38%)

The reactions were conducted as described in Table 2
[a] Isolated yields,
[b] Before addition of **1a**, a mixture of **2** and **3** in DMF was stirred at rt for 1 h in the presence of CuCl,
[c] One-portion Mannich-type reaction of **1a**, **2**, and **3d** was conducted at 100 °C for 1 h

relatively low reactivity to give **12g** and **12 h** in 30 and 38% respective yields (entries 2 and 3).

Finally, the author examined preparation of 3-(aminomethyl)isoquinolines fused with various heterocycles, by changing the carbon tether of the diamine component **8** (Table 4). Use of 1,2-diaminoethane **8b** in the reaction of 2-ethynylbenzaldehyde **1a**, paraformaldehyde **2** and diisopropylamine **3a** in the presence of CuCl under an oxygen atmosphere gave the desired 2,3-dihydroimidazo[2,1-a]isoquinoline **13** in 56% yield (entry 1). The reaction using 1,4-diaminobutane **8c** afforded the tricyclic compound **14** with a tetrahydro[1,3]diazepine structure in 50% yield (entry 3). The limitation of this reaction can be seen in the reaction with 1,5-diaminopentane **8d**, which produced 1,3-diazocine-fused isoquinoline **15** in only 12% yield (entry 5). This strategy was also applicable to the synthesis of tetracyclic benzimidazo[2,1-a]isoquinoline **16** (entry 7) [5].[a] In the case of entries 4 and 8, the increased yields of **14** and **16** were observed under an argon atmosphere, although a prolonged reaction time was required (15 h for the cyclization/oxidation step).

In conclusion, the author has developed a novel route to isoquinoline-fused polycyclic compounds by a four-component coupling and cascade cyclization strategy. In this reaction, the cyclization/oxidation step can be accelerated by use of an oxygen atmosphere, giving rise to improved yields of the cyclized products in many cases. Because this four-component reaction catalytically forms one carbon–carbon and four carbon–nitrogen bonds producing only H_2O and H_2 as the

Table 4 Synthesis of (Aminomethyl)isoquinoline-fused polycyclic compounds

Entry	Diamine	Atmosphere[a]	Product (yield)[b]
	H_2N~~~NH_2 (8b structure)		(product structure with $N(i\text{-}Pr)_2$)
1	8b	O_2	13 (56%)
2	8b	Ar	13 (53%)
	H_2N~~~NH_2 (8c structure)		(product structure with $N(i\text{-}Pr)_2$)
3	8c	O_2	14 (50%)
4	8c	Ar	14 (63%)
	H_2N~~~NH_2 (8d structure)		(product structure with $N(i\text{-}Pr)_2$)
5	8d	O_2	15 (12%)
6	8d	Ar	15 (5%)
	(8e structure, NH_2, NH_2)		(product structure with $N(i\text{-}Pr)_2$)
7	8e	O_2	16 (44%)
8	8e	Ar	16 (58%)

The reactions were conducted as described in Table 2
[a] The reaction under argon required 15 h for the cyclization/oxidation step,
[b] Isolated yields

theoretical waste products, it would be useful for diversity oriented synthesis of various isoquinolines in an atom-economical manner.

6.1 Experimental Section

6.1.1 General Procedure for Synthesis of (Aminomethyl)isoquinoline-Fused Polycyclic Compounds by Domino Mannich-Type Reaction and Cascade Cyclization: Synthesis of 6-[(N,N-Diisopropylamino)methyl]-3,4-dihydro-2H-pyrimido[2,1-a] isoquinoline (12a) (Table 1, Entry 10)

A mixture of 2-ethynylbenzaldehyde **1a** (25.0 mg, 0.19 mmol), paraformaldehyde **2** (11.5 mg, 0.38 mmol), diisopropylamine **3a** (53.8 µL, 0.38 mmol) and CuCl

(1.9 mg, 0.019 mmol) in DMF (1.5 mL) was stirred under O_2 at rt for 1.5 h. After the Mannich-type reaction was completed monitored by TLC, propanediamine **8a** (48.1 μL, 0.58 mmol) and MS 4 Å (37.5 mg) were added and the mixture was additionally stirred at 120 °C for 1 h. The mixture was concentrated in vacuo and purified by column chromatography over alumina with $CHCl_3/CH_3OH$ (15:1) as the eluent to give **12a** (41.3 mg 72%) as a solid mass: mp 128–129 °C; ^{1}H NMR (400 MHz, CDCl$_3$) δ 1.04 (d, $J = 6.6$ Hz, 12H, 4 × CH$_3$), 1.91–1.96 (m, 2H, 3-CH$_2$), 3.06–3.16 (m, 2H, 2 × CH(CH$_3$)$_2$), 3.49 (s, 2H, NCH$_2$), 3.64 (t, $J = 5.6$ Hz, 2H, NCH$_2$), 4.13 (t, $J = 5.9$ Hz, 2H, NCH$_2$), 6.05 (s, 1H, 7-H), 7.19 (d, $J = 7.8$ Hz, 1H, Ar), 7.23–7.27 (m, 1H, Ar), 7.36–7.40 (m, 1H, Ar), 8.26 (d, $J = 8.0$ Hz, 1H, Ar); ^{13}C NMR (100 MHz, CDCl$_3$) δ 20.3 (4C), 21.0, 43.5, 44.4, 47.2 (2C), 48.0, 105.3, 124.9, 125.6, 126.1, 127.2, 130.2, 134.0, 140.7, 149.9; MS (FAB) m/z (%): 298 (MH$^+$, 100); HRMS (FAB) calcd for $C_{19}H_{28}N_3$ (MH$^+$): 298.2284; found: 298.2285.

6.1.2 6-[(N,N-Diisopropylamino)methyl]-9-Fluoro-3,4-Dihydro-2H-Pyrimido[2,1-a]isoquinoline (12b)

By a procedure identical to that described for **12a** from **1a**, **1b** (28.5 mg, 0.19 mmol) was converted into **12b** (33.3 mg, 55%) as a pale yellow solid: mp 123–125 °C; ^{1}H NMR (400 MHz, CDCl$_3$) δ 1.04 (d, $J = 6.6$ Hz, 12H, 4 × CH$_3$), 1.90–1.96 (m, 2H, 3-CH$_2$), 3.05–3.15 (m, 2H, 2 × CH(CH$_3$)$_2$), 3.47 (s, 2H, NCH$_2$), 3.62 (t, $J = 5.6$ Hz, 2H, NCH$_2$), 4.11 (t, $J = 5.9$ Hz, 2H, NCH$_2$), 5.99 (s, 1H, 7-H), 6.82 (dd, $J = 9.4, 2.6$ Hz, 1H, Ar), 6.90–6.95 (m, 1H, Ar), 8.24 (dd, $J = 8.9, 6.0$ Hz, 1H, Ar); ^{13}C NMR (100 MHz, CDCl$_3$) δ 20.3 (4C), 21.0, 43.5, 44.4, 47.4 (2C), 47.9, 104.1, 109.8 (d, $J = 21.5$ Hz), 113.8 (d, $J = 22.3$ Hz), 123.8, 128.4, (d, $J = 9.1$ Hz), 136.1 (d, $J = 9.9$ Hz), 142.4, 149.1, 164.2 (d, $J = 248.3$ Hz); MS (FAB) m/z (%): 316 (MH$^+$, 100); HRMS (FAB) calcd for $C_{19}H_{27}FN_3$ (MH$^+$): 316.2189; found: 316.2188.

6.1.3 6-[(N,N-Diisopropylamino)methyl]-10-Fluoro-3,4-Dihydro-2H-Pyrimido[2,1-a]isoquinoline (12c).

By a procedure identical to that described for **12a** from **1a**, **1c** (28.5 mg, 0.19 mmol) was converted into **12c** (44.6 mg, 74%) as a pale yellow solid: mp 139–141 °C; ^{1}H NMR (400 MHz, CDCl$_3$) δ 1.03 (d, $J = 6.6$ Hz, 12H, 4 × CH$_3$), 1.90–1.95 (m, 2H, 3-CH$_2$), 3.05–3.15 (m, 2H, 2 × CH(CH$_3$)$_2$), 3.47 (s, 2H, NCH$_2$), 3.63 (t, $J = 5.5$ Hz, 2H, NCH$_2$), 4.12 (t, $J = 5.7$ Hz, 2H, NCH$_2$), 6.01 (s, 1H, 7-H), 7.07–7.18 (m, 2H, Ar), 8.24 (dd, $J = 10.6, 2.6$ Hz, 1H, Ar); ^{13}C NMR (100 MHz, CDCl$_3$) δ 20.2 (4C), 20.9, 43.3, 44.5, 47.2 (2C), 47.8, 103.9, 111.1 (d,

$J = 23.2$ Hz),   118.1   (d, $J = 23.2$ Hz),   126.7   (d, $J = 7.4$ Hz),   129.2,   (d, $J = 8.3$ Hz), 130.4 (d, $J = 2.5$ Hz), 140.0 (d, $J = 2.5$ Hz), 148.9 (d, $J = 3.3$ Hz), 161.3 (d, $J = 244.1$ Hz); MS (FAB) m/z (%): 316 (MH$^+$, 100); HRMS (FAB) calcd for $C_{19}H_{27}FN_3$ (MH$^+$): 316.2189; found: 316.2180.

6.1.4 6-[(N,N-Diisopropylamino)methyl]-9-Methyl-3,4-Dihydro-2H-Pyrimido[2,1-a]isoquinoline (12d)

By a procedure identical to that described for **12a** from **1a**, **1d** (27.7 mg, 0.19 mmol) was converted into **12d** (42.2 mg, 71%) as a pale yellow solid: mp 132–135 °C; ^{1}H NMR (400 MHz, CDCl$_3$) δ 1.03 (d, $J = 6.6$ Hz, 12H, 4 × CH$_3$), 1.90–1.95 (m, 2H, 3-CH$_2$), 2.36 (s, 3H, ArCH$_3$), 3.05–3.15 (m, 2H, 2 × CH(CH$_3$)$_2$), 3.46 (s, 2H, NCH$_2$), 3.63 (t, $J = 5.6$ Hz, 2H, NCH$_2$), 4.11 (t, $J = 5.9$ Hz, 2H, NCH$_2$), 5.99 (s, 1H, 7-H), 6.98–7.00 (m, 1H, Ar), 7.08 (dd, $J = 8.3$, 1.5 Hz, 1H, Ar), 8.15 (d, $J = 8.3$ Hz, 1H, Ar); ^{13}C NMR (100 MHz, CDCl$_3$) δ 20.2 (4C), 20.9, 21.4, 43.4, 44.2, 47.1 (2C), 47.9, 105.4, 124.6, 124.9, 125.6, 127.6, 134.0, 140.4, 140.6, 150.0; MS (FAB) m/z (%): 312 (MH$^+$, 100); HRMS (FAB) calcd for $C_{20}H_{30}N_3$ (MH$^+$): 312.2440; found: 312.2443.

6.1.5 6-[(N,N-Diisopropylamino)methyl]-10-Methoxy-3,4-Dihydro-2H-Pyrimido[2,1-a]isoquinoline (12e)

By a procedure identical to that described for **12a** from **1a**, **1e** (30.8 mg, 0.19 mmol) was converted into **12e** (34.8 mg, 55%) as a pale yellow solid: mp 174–176 °C; ^{1}H NMR (400 MHz, CDCl$_3$) δ 1.03 (d, $J = 6.6$ Hz, 12H, 4 × CH$_3$), 1.92–1.97 (m, 2H, 3-CH$_2$), 3.05–3.15 (m, 2H, 2 × CH(CH$_3$)$_2$), 3.49 (s, 2H, NCH$_2$), 3.67 (t, $J = 5.5$ Hz, 2H, NCH$_2$), 3.90 (s, 3H, OMe), 4.16 (t, $J = 5.7$ Hz, 2H, NCH$_2$), 6.02 (s, 1H, 7-H), 7.02 (dd, $J = 8.5$, 2.7 Hz, 1H, Ar), 7.14 (d, $J = 8.5$ Hz, 1H, Ar), 7.75–7.77 (m, 1H, Ar); ^{13}C NMR (100 MHz, CDCl$_3$) δ 20.2 (4C), 21.0, 43.5, 45.6, 47.0 (2C), 47.9, 55.6, 105.0, 106.2, 120.3, 126.6, 127.8, 128.4, 138.2, 149.8, 158.4; MS (FAB) m/z (%): 328 (MH$^+$,100); HRMS (FAB) calcd for $C_{20}H_{30}ON_3$ (MH$^+$): 328.2389; found: 328.2383.

6.1.6 6-(Piperidin-1-ylmethyl)-3,4-Dihydro-2H-Pyrimido[2,1-a]isoquinoline (12f)

A mixture of paraformaldehyde **2** (17.3 mg, 0.58 mmol), piperidine **3b** (57.0 μL, 0.58 mmol) and CuCl (1.9 mg, 0.019 mmol) in DMF (1.5 mL) was stirred under O$_2$ at rt for 1 h. Then 2-ethynylbenzaldehyde **1a** (25.0 mg, 0.19 mmol) was added at rt, and the mixture was additionally stirred at this temperature for 1.5 h. After the Mannich-type reaction was completed monitored by TLC, propanediamine **8a** (48.1 mL, 0.58 mmol) and MS 4 Å (37.5 mg) were added and the mixture was stirred at 120 °C for 1 h. The mixture was concentrated in vacuo and purified by column chromatography over alumina with $CHCl_3/CH_3OH$ (20:1) as the eluent to give **12f** (41.3 mg 61%) as a brown oil: ^1H NMR (400 MHz, $CDCl_3$) δ 1.40–1.47 (m, 2H, CH_2), 1.52–1.57 (m, 4H, 2 × CH_2), 1.92–1.98 (m, 2H, 3-CH_2), 2.35–2.43 (m, 4H, 2 × CH_2), 3.19 (s, 2H, NCH_2), 3.65 (t, $J = 5.6$ Hz, 2H, NCH_2), 4.09 (t, $J = 5.9$ Hz, 2H, NCH_2), 5.87 (s, 1H, 7-H), 7.17 (d, $J = 7.6$ Hz, 1H, Ar) 7.23–7.27 (m, 1H, Ar), 7.35–7.39 (m, 1H, Ar), 8.25 (d, $J = 8.0$ Hz, 1H, Ar); ^{13}C NMR (100 MHz, $CDCl_3$) δ 21.1, 24.3, 26.1 (2C), 43.7, 44.6, 54.1 (2C), 61.5, 105.2, 124.9, 125.5, 126.1, 127.7, 130.1, 133.7, 138.7, 149.7; MS (FAB) m/z (%): 282 (MH^+, 100); HRMS (FAB) calcd for $C_{18}H_{24}N_3$ (MH^+): 282.1970; found: 282.1974.

6.1.7 6-[(N,N-Diallylamino)methyl]-3,4-Dihydro-2H-Pyrimi-do[2,1-a]isoquinoline (12g)

By a procedure similar to that described for **12a** from **1a**, **1a** (25.0 mg, 0.19 mmol) was converted into **12g** (17.0 mg, 30%) using diallylamine **3c** (71.1 μL, 0.58 mmol): brown oil; ^1H NMR (400 MHz, $CDCl_3$) δ 1.92–1.98 (m, 2H, 3-CH_2), 3.10–3.12 (m, 4H, 2 × NCH_2), 3.34 (s, 2H, NCH_2), 3.65 (t, $J = 5.5$ Hz, 2H, NCH_2), 4.07 (t, $J = 5.9$ Hz, 2H, NCH_2), 5.16–5.21 (m, 4H, 2 × C = CH_2), 5.79–5.89 (m, 2H, 2 × C = CH), 5.94 (s, 1H, 7-H), 7.18 (d, $J = 7.6$ Hz, 1H, Ar) 7.25–7.29 (m, 1H, Ar), 7.36–7.41 (m, 1H, Ar), 8.26 (d, $J = 8.0$ Hz, 1H, Ar); ^{13}C NMR (100 MHz, $CDCl_3$) δ 21.0, 44.0, 44.3, 56.0, 56.2 (2C), 106.2, 118.2 (2C), 125.0, 125.7, 126.5, 127.3, 130.4, 133.7, 134.9 (2C), 138.9, 149.8; MS (FAB) m/z (%): 294 (MH^+, 100); HRMS (FAB) calcd for $C_{19}H_{24}N_3$ (MH^+): 294.1970; found: 294.1969.

6.1.8 6-{[N,N-Bis((R)-1-phenylethyl)amino]methyl}-3,4-Dihydro-2H-Pyrimido[2,1-a]isoquinoline (12h)

By a procedure similar to that described for **12a** from **1a**, **1a** (25 mg, 0.19 mmol) was converted into **12h** (30.9 mg, 38%) using bis[(R)-1-phenylethyl]amine **3d** (87.9 μL, 0.38 mmol): colorless solid; mp 174–176 °C; ^1H NMR (400 MHz, $CDCl_3$) δ 1.42–1.68 (m, 8H, 3-CH_2 and 2 × CH_3), 2.77–2.83 (m, 1H, NCH), 3.43–3.58 (m, 5H, NCH and 2 × NCH_2), 4.17 (q, $J = 6.9$ Hz, 2H, 2 × CH_3CH), 6.05 (s, 1H, 7-H), 7.11–7.38 (m, 13H, Ar), 8.19 (d, $J = 8.0$ Hz, 1H, Ar); ^{13}C NMR (100 MHz, $CDCl_3$) δ 14.5 (2C), 20.8, 42.6, 44.3, 47.8, 55.1 (2C), 106.0, 124.8, 125.5, 126.1, 126.8 (2C), 127.6, 127.8 (4C), 128.1 (4C), 130.1, 133.6, 139.9, 143.5

(2C), 149.5; MS (FAB) m/z (%): 422 (MH$^+$, 100); HRMS (FAB) calcd for $C_{29}H_{32}N_3$ (MH$^+$): 422.2596; found: 422.2602.

6.1.9 5-[(N,N-Diisopropylamino)methyl]-2,3-Dihydroimidazo[2,1-a]isoquinoline (13)

By a procedure similar to that described for **12a** from **1a**, **1a** (25.0 mg, 0.19 mmol) was converted into **13** (30.5 mg, 56%) using ethylenediamine **8b** (38.7 μL, 0.58 mmol): brown oil; ^{1}H NMR (400 MHz, CDCl$_3$) δ 1.04 (d, J = 6.6 Hz, 12H, 4 × CH$_3$), 3.06–3.16 (m, 2H, 2 × CHCH$_3$), 3.43 (s, 2H, NCH$_2$), 4.03–4.09 (m, 2H, NCH$_2$), 4.20–4.26 (m, 2H, NCH$_2$), 5.97 (s, 1H, 7-H), 7.24–7.27 (m, 2H, Ar), 7.41–7.45 (m, 1H, Ar), 8.10 (d, J = 8.3 Hz, 1H, Ar); ^{13}C NMR (100 MHz, CDCl$_3$) δ 20.3 (4C), 47.2, 47.50, 47.54 (2C), 53.1, 102.8, 121.7, 125.2, 125.6, 126.0, 131.2, 136.5, 140.9, 158.6; MS (FAB) m/z (%): 284 (MH$^+$, 100); HRMS (FAB) calcd for $C_{18}H_{26}N_3$ (MH$^+$): 284.2127; found: 284.2134.

6.1.10 7-[(N,N-Diisopropylamino)methyl]-2,3,4,5-Tetrahy-dro[1,3]diazepino[2,1-a]isoquinoline (14)

By a procedure similar to that described for **12a** from **1a**, **1a** (25.0 mg, 0.19 mmol) was converted into **14** (29.8 mg, 63%) using butanediamine **8c** (57.9 μL, 0.38 mmol) under argon: brown oil; ^{1}H NMR (400 MHz, CDCl$_3$) δ 1.05 (d, J = 6.8 Hz, 12H, 4 × CH$_3$), 1.93–1.99 (m, 2H, CH$_2$), 2.03–2.09 (m, 2H, CH$_2$), 3.09–3.19 (m, 2H, 2 × CHCH$_3$), 3.50 (s, 2H, NCH$_2$), 3.89–3.92 (m, 2H, NCH$_2$), 4.03–4.06 (m, 2H, NCH$_2$), 6.12 (s, 1H, 8-H), 7.16 (d, J = 7.6 Hz, 1H, Ar), 7.22–7.26 (m, 1H, Ar), 7.33–7.37 (m, 1H, Ar), 8.16 (d, J = 7.8 Hz, 1H, Ar); ^{13}C NMR (100 MHz, CDCl$_3$) δ 20.2 (4C), 25.4, 26.6, 47.2 (2C), 47.7, 47.8, 48.2, 105.9, 124.4, 125.6, 126.0, 128.5, 129.8, 134.1 142.6, 153.9; MS (FAB) m/z (%): 312 (MH$^+$, 100); HRMS (FAB) calcd for $C_{20}H_{30}N_3$ (MH$^+$): 312.2440; found: 312.2433.

6.1.11 8-[(N,N-Diisopropylamino)methyl]-3,4,5,6-Tetrahydro-2H-[1,3]diazocino[2,1-a]isoquinoline (15)

By a procedure similar to that described for **12a** from **1a**, **1a** (25.0 mg, 0.19 mmol) was converted into **15** (7.2 mg, 12%) using pentanediamine **8d** (67.8 μL, 0.38 mmol): brown oil; ^{1}H NMR (400 MHz, CDCl$_3$) δ 1.03 (d, J = 6.6 Hz, 12H, 4 × CH$_3$), 1.63–1.69 (m, 2H, CH$_2$), 1.91–2.04 (m, 4H, 2 × CH$_2$), 3.08–3.18 (m,

2H, 2 × C*H*CH₃), 3.44 (s, 2H, NCH₂), 4.20 (t, $J = 6.2$ Hz, 2H, NCH₂), 4.43 (t, $J = 6.6$ Hz, 2H, NCH₂), 6.16 (s, 1H, 9-H), 7.17 (d, $J = 7.6$ Hz, 1H, Ar), 7.23–7.26 (m, 1H, Ar), 7.34–7.38 (m, 1H, Ar), 8.31 (d, $J = 8.0$ Hz, 1H, Ar); ¹³C NMR (100 MHz, CDCl₃) δ 20.2 (4C), 20.6, 29.2, 30.7, 45.4, 46.6, 47.2 (2C), 47.6, 106.3, 124.5, 126.1, 126.9, 129.4, 129.8, 134.2, 141.7, 150.2; MS (FAB) *m/z* (%): 326 (MH⁺, 100); HRMS (FAB) calcd for C₂₁H₃₂N₃ (MH⁺): 326.2596; found: 326.2597.

6.1.12 6-[(N,N-Diisopropylamino)methyl]benzimidazo[2,1-a]iso-quinoline (16)

By a procedure similar to that described for **12a** from **1a**, **1a** (25.0 mg, 0.19 mmol) was converted into **16** (28.1 mg, 58%) using phenylendiamine **8e** (62.3 μL, 0.38 mmol) under argon: pale yellow solid; mp 152–154 °C; ¹H NMR (400 MHz, CDCl₃) δ 1.16 (d, $J = 6.3$ Hz, 12H, 4 × CH₃), 3.22–3.32 (m, 2H, 2 × C*H*CH₃), 4.36 (d, $J = 1.2$ Hz, 2H, NCH₂), 7.36–7.40 (m, 1H, Ar), 7.49–7.53 (m, 2H, Ar and 5-H), 7.60–7.68 (m, 2H, Ar), 7.74 (d, $J = 7.3$ Hz, 1H, Ar), 8.06 (d, $J = 8.0$ Hz, 1H, Ar), 8.14 (d, $J = 8.3$ Hz, 1H, Ar), 8.84–8.86 (m, 1H, Ar); ¹³C NMR (100 MHz, CDCl₃) δ 20.9 (4C), 47.6, 49.5 (2C), 109.0, 114.7, 119.9, 121.4, 122.2, 124.0, 125.0, 126.3, 127.0, 129.8, 130.9, 131.8, 140.9, 144.3, 148.6; MS (FAB) *m/z* (%): 332 (MH⁺, 100); HRMS (FAB) calcd for C₂₂H₂₆N₃ (MH⁺): 332.2127; found: 332.2133.

References

1. Handley DA, Van Valen RG, Melden MK, Houlihan WJ, Saunders RN (1988) J Pharmacol Exp Ther 247:617–623
2. Houlihan WJ, Cheon SH, Parrino VA, Handley DA, Larson DA (1993) J Med Chem 36:3098–3102
3. Scholz D, Schmidt H, Prieschl EE, Csonga R, Scheirer W, Weber V, Lembachner A, Seidl G, Werner G, Mayer P, Baumruker T (1998) J Med Chem 41:1050–1059
4. Griffin RJ, Fontana G, Golding BT, Guiard S, Hardcastle IR, Leahy JJJ, Martin N, Richardson C, Rigoreau L, Stockley M, Smith GCM (2005) J Med Chem 48:569–585
5. Danhauser-Riedl S, Felix SB, Houlihan WJ, Zafferani M, Steinhauser G, Oberberg D, Kalvelage H, Busch R, Rastetter J, Berdel WE (1991) Cancer Res 51:43–48
6. Houlihan WJ, Munder PG, Handley DA, Cheon SH, Parrino VA (1995) J Med Chem 38:234–240
7. Parenty ADC, Smith LV, Guthrie KM, Long D-L, Plumb J, Brown R, Cronin L (2005) J Med Chem 48:4504–4506
8. Smith LV, Parenty ADC, Guthrie KM, Plumb J, Brown R, Cronin L (2006) ChemBioChem 7:1757–1763
9. Chaykovsky M, Benjamin L, Ian Fryer R, Metlesics WJ (1970) J Org Chem 35:1178–1180
10. Houlihan WJ, Parrino VA (1982) J Org Chem 47:5177–5180
11. Loones KTJ, Maes BUW, Dommisse RA, Lemière GLF (2004) Chem Commun 2466–2467

12. Parenty ADC, Smith LV, Pickering AL, Long D-L, Cronin L (2004) J Org Chem 69:5934–5946
13. Sharon A, Pratap R, Maulik PR, Ram VJ (2005) Tetrahedron 61:3781–3787
14. Kiselyov AS (2005) Tetrahedron Lett 46:4487–4490
15. Parenty, A. D. C.; Guthrie, K. M.; Song, Y.-F.; Smith, L. V.; Burkholder, E.; Cronin, L. *Chem. Commun.* **2006**, 1194–1196
16. Loones KTJ, Maes BUW, Herrebout WA, Dommisse RA, Lemière GLF, Van der Veken BJ (2007) Tetrahedron 63:3818–3825
17. Hubbard JW, Piegols AM, Söderberg BCG (2007) Tetrahedron 63:7077–7085
18. Parenty ADC, Cronin L (2008) Synthesis 155–160
19. Ohta Y, Oishi S, Fujii N, Ohno H (2008) Chem Commun 835–837
20. Dyker G, Stirner W, Henkel G (2000) Eur J Org Chem 1433–1441
21. Su S, Porco JA Jr (2007) J Am Chem Soc 129:7744–7745
22. Asao N, Iso K, Yudha SS (2006) Org Lett 8:4149–4151
23. Ding Q, Wang B, Wu J (2007) Tetrahedron 63:12166–12171
24. Ding Q, Wu J (2007) Org Lett 9:4959–4962
25. Gao K, Wu J (2007) J Org Chem 72:8611–8613
26. Ye Y, Ding Q, Wu J (2008) Tetrahedron 64:1378–1382
27. Ohtaka M, Nakamura H, Yamamoto Y (2004) Tetrahedron Lett 45:7339–7341
28. Asao N, Yudha SS, Nogami T, Yamamoto Y (2005) Angew Chem Int Ed 44:5526–5528
29. Yanada R, Obika S, Kono H, Takemoto Y (2006) Angew Chem Int Ed 45:3822–3825
30. Obika S, Kono H, Yasui Y, Yanada R, Takemoto Y (2007) J Org Chem 72:4462–4468

Chapter 7
Conclusions

1. Copper-catalyzed synthesis of 2-(aminomethyl)indole by domino three-component coupling–cyclization was accomplished. This reaction proceeds through Mannich-type reaction using 2-ethynylanilines, aldehydes, and secondary amines, followed by hydroamination. This is the first example of three-component indole formation without producing any salts as a byproduct. Using alkyl aldehydes and the chiral ligand PINAP, the corresponding indole bearing a branched substituent was produced with moderate ee values. This indole formation was applicable to the synthesis of indole-fused polycyclic compounds via palladium-catalyzed C–H functionalization at 3-position of indole. Synthetic application to calindol, benzo[e][1,2]thiazines, and indene was also conducted.

2. β-Carboline structure was constructed by one-pot reaction, which involves the three-component indole formation and nucleophilic cyclization by the addition of t-BuOK or MsOH. This is the first example of multi-component synthesis of carbolines, except for those using the Pictet-Spengler type reaction. Utilizing the three-component indole formation, indole-fused 1,4-diazepines were also synthesized through deprotection/N-arylation at nitrogen atom of indole by one-pot addition of MeONa after the formation of indole. These reactions form four bonds in a single reaction vessel, which involves two C–C bonds/two C–N bonds or one C–C bond/three C–N bonds.

3. In relation to the three-component indole formation, a novel four-component synthesis of 3-(aminomethyl)isoquinoline was developed. The reaction of 2-ethynylbenzaldehyde with $(HCHO)_n$, secondary amine, and t-BuNH$_2$ proceeds through Mannich-type reaction, cyclization, and elimination of t-butyl group. By the use of alkane diamine instead of t-BuNH$_2$, 3-(aminomethyl)isoquinoline-fused polycyclic compounds were also synthesized by cascade cyclization and oxidation. Changing the carbon tether of the diamine component led to the synthesis of isoquinolines fused with various heterocycles.

Taken together, the author has achieved the development for the copper-catalyzed synthesis of 2-(aminomethyl)indoles and 3-(aminomethtyl)isoquinolines

Y. Ohta, *Copper-Catalyzed Multi-Component Reactions*, Springer Theses,
DOI: 10.1007/978-3-642-15473-7_7, © Springer-Verlag Berlin Heidelberg 2011

by catalytic domino reaction including multi-component coupling. These findings would contribute to the diversity-oriented synthesis for the drug discovery and facile synthesis of biologically active natural products containing complex structure. Futhermore, indole- or isoquinoline-fused polycyclic compounds were also synthesized through this multi-component reaction and one-pot addition of acid or base. These investigations may provide the development for the synthesis of bioactive compounds in an atom-economical manner, which could lead to development of promising drug leads with structural complexity.